STATISTICAL MECHANICS FOR BEGINNERS

A Textbook for Undergraduates

Second Edition

STATISTICAL MECHANICS FOR BEGINNERS

A Textbook for Undergraduates

Second Edition

LUCIEN GILLES BENGUIGUI
Technion-Israel Institute of Technology, Israel

World Scientific

NEW JERSEY • LONDON • SINGAPORE • BEIJING • SHANGHAI • TAIPEI • CHENNAI

Published by

World Scientific Publishing Co. Pte. Ltd.
5 Toh Tuck Link, Singapore 596224
USA office: 27 Warren Street, Suite 401-402, Hackensack, NJ 07601
UK office: 57 Shelton Street, Covent Garden, London WC2H 9HE

Library of Congress Control Number: 2025022976

British Library Cataloguing-in-Publication Data
A catalogue record for this book is available from the British Library.

STATISTICAL MECHANICS FOR BEGINNERS
A Textbook for Undergraduates
Second Edition

ISBN 978-981-98-1434-3 (hardcover)
ISBN 978-981-98-1486-2 (paperback)
ISBN 978-981-98-1435-0 (ebook for institutions)
ISBN 978-981-98-1436-7 (ebook for individuals)

For any available supplementary material, please visit
https://www.worldscientific.com/worldscibooks/10.1142/14344#t=suppl

Desk Editors: Murali Appadurai Joseph Ang

Typeset by Stallion Press
Email: enquiries@stallionpress.com

Preface of the Second Edition

In writing this second edition of this book *Statistical Mechanics for Beginners*, I intended to keep the book for students who for the first time study this scientific field. The book may be also useful for teachers in proposing a frame for a course. I think that this approach makes the originality of the book. The question was: what to do in a second edition? This field was initiated more than 100 years ago and one cannot hope for recent development which could be added to a first edition. So I decided to read the book as it was a new one. In reading I ask myself: is this text clear enough? Are the explanations enough satisfying? The first thing to do was to correct the text from inevitable errors. And I made several minor corrections to all the text. But more important are four initiatives.

The first concerns the fluctuations. Since the theory gives distributions, it is good to recall this point that I mention rapidly without giving too many details.

In the second initiative, I wanted to present a detailed study of the bosons. The goal was to understand more clearly the Bose–Einstein condensation. I confess that, in the first edition, this phenomenon did not receive a satisfactorily explanation. The answer to the problem did not seem to correspond to the question. I consider a group of bosons with a given potential and calculate numerically their properties: energy, chemical potential and variation of the population in the lowest level with the temperature. This last point has great importance because this is a particular behaviour of bosons. When one decreases the temperature, the ground level is progressively filled. However, one distinguishes two regions, one with the lowest level almost empty and the second region with a filling of the lowest level in a small temperature interval. This is like the Bose–Einstein condensation but progressively. The Bose–Einstein condensation appears, as in a phase transition, at a well-defined temperature but otherwise, there is a blurred transition. A Bose–Einstein condensation is not a sort of exotic phenomenon but an intrinsic property of bosons. In the case of fermions, in an identical situation, the ground level is filled regularly when the temperature is decreased.

But the story is not finished. Why the problem of the chemical potential is solved by the division of the particles? Why it is proposed to say that the series and the integral do not agree, in this case? In fact it is the general problem to compare the sum of a series and that of the integral, both with the same expression. It is not evidence that one can replace a series by an integral as physicists make currently. Series and integrals have very different definitions, they are very different mathematical objects. The problem of the comparison of the sums of series and integrals was solved by Euler and McLaurin in 1737. The Euler–McLaurin formula gives the difference between the sums, but it seems that the Bose–Einstein condensation is one case where the difference is important. It is likely that in most cases where one replaces a series by an integral, neglecting the differences has no serious consequences.

The third initiative is the energy of a gas of fermions. I give two solutions to the problem of calculating its energy. The second solution that I added is much known and it is simpler than the first. But I want to show that the first solution, in spite of its complication has the advantage of giving a useful formula.

Finally, I added the problem of the size of a neutron star. This is a kind of an introduction to astrophysics.

I mention that the number of exercises is much larger than in the first edition and a large part is new and not published elsewhere.

I hope these four initiatives will contribute to make the book more useful for beginners.

L.G. Benguigui
Haifa 2024

Preface of the First Edition

This book is intended for the students who begin for the first time the study of statistical mechanics. There are two different approaches to teach thermal physics to the students in physics at the level of the BSc. The first is to expose the subject in two separate courses: one in thermodynamics i.e. the macroscopic aspect of thermal physics and a second one (taught immediately after) in statistical mechanics, i.e. the microscopic aspect. There are excellent books that chose this way. The other approach is to present the subject in only one course mixing the two aspects. Excellent books also exist that follow this more compact method. Here there is no place to discuss the advantages and the disadvantages of each of the two approaches. In this book, I follow the first one. This means that this course in suited for students that have already some knowledge in thermodynamics.

Historically, the classical statistical mechanics was first developed and only later, with the progress of the theory of the quantum theory, the quantum statistical mechanics was born. I think that, from a pedagogical point of view, it is easier to teach the quantum statistical mechanics than its classical counter part. This is the reason why in this book, the major part of the book is devoted to the quantum statistical mechanics. It suffices that the student has an elementary knowledge of the basic results of the quantum theory to be able to understand the matter.

This book may appear very short. It is effectively far form being complete. For example, in thermodynamics the concept of entropy is introduced in connection with irreversible process. However, in this book, I did not discuss this problem, giving time for student to study it later.

I tried to present the subject in a consistent form: first the general principles or the methods giving the links between the macroscopic and the microscopic worlds. In addition, in the second part, applications to simple situations are developed. It is good to give first the foundations and only after the details of the applications. On the other side, I present classical cases as particular situations of quantum cases. This is not the way in which the matter is frequently taught. I think that the actual presentation has some novel aspect. The mathematical level is not very high. The reader has to be used with algebraic calculus, combinatory, differential and integral calculus.

The book is almost exclusively for the use of the students. It is based on my personal teaching at the Technion. At the disposal of the teacher, there are many very good books with a lot of complementary details for an oral teaching in the classroom. But I did not find a book that I can recommend to the students when, for example, he was not able to assist some classes. When I taught this course, I had only two hours a week (and one hour for exercises) during one semester of 14 weeks. In such limited time, only the main points may be taught. This means that all is important in the book. It represents what a student needs to know in order to be able to follow other courses in his studies toward his first degree in physics (for example, a course in solid state physics). I introduced exercises which are straightforward applications of the matter of each chapter. They will help the student to assimilate the main concepts and methods.

I added a special chapter on the history of statistical mechanics. Since in the book itself I do not follow the historical development, I thought this could be interesting to bring some views about how the theory was built.

I thank J. Unffick of University of Utrecht for his help in preparing the chapter on the history of the statistical mechanics.

L.G. Benguigui
Haifa 2009

About the Author

L.G. Benguigui received his PhD from the University of Paris, he began his career at the Technion the Israel Institute of Technology until his retirement. His research activity was directed towards various aspects of condensed matter: phase transition, liquid crystals and polymers, and disordered systems (co author with B.K. Chakrabarty). He was among the first physicists to use physical methods (fractals, networks) for studying human geography. The recent works of L.G. Benguigui, concern the history of physics, He published three books: one of statistical physics of disordered systems, another on the history of the Relativity theory and the first edition of the present book.

L.G. Benguigui has also literary activity. He published (in French) a book on the well known author of tragedies Racine and another nook on the Jewish youth in France after the second world war.

Contents

Physical Constants

Boltzmann constant	$k_B = 1.3896 \times 10^{-23}\,\mathrm{J/K}$
	$k_B = 8.517 \times 10^{-5}\,\mathrm{eV/K}$
Planck constant	$h = 6.62607 \times 10^{-34}\,\mathrm{J\,s}$
	$\hbar = h/2\pi = 1.0546 \times 10^{-34}\,\mathrm{J\,s}$
Electron mass	$m = 9.1094 \times 10^{-31}\,\mathrm{kg}$
Proton mass	$M = 1.6726 \times 10^{-27}\,\mathrm{kg}$
Electric charge of electron	$e = 1.6021 \times 10^{-19}\,\mathrm{C}$
Avogadro number	$N_A = 6.022 \times 10^{23}$
Speed of light	$c = 2.9979 \times 10^{8}\,\mathrm{m/s}$
Gas constant	$R = N_A k_B = 8.3145\,\mathrm{J/K}$
Electron volt	$\mathrm{eV} = 1.6021 \times 10^{-19}\,\mathrm{J}$

Introduction

In thermodynamics, it is shown that the thermal properties of a system compound of a very large number of particles is characterized by a relatively small number of quantities such as the internal energy, the temperature, the entropy, the volume, and the pressure. These are the macroscopic parameters of the matter. Thermodynamics was developed without a hypothesis about a microscopic picture of the matter in its three forms: solid, liquid or gas. However, with the development of the atomic theory, it has become possible to look for the link between the macroscopic world and a microscopic picture. At the end of the 19th and at the beginning of the 20th century, the first steps towards a theory relating the macroscopic world and a microscopic picture were proposed by Boltzmann and Gibbs. At this time the term Statistical Mechanics was coined by Gibbs.

The basic problem in statistical mechanics is to find the macroscopic properties of a system of particles from the knowledge of their microscopic properties. But at the microscopic level, the number of parameters is enormous. It is impossible to follow each particle individually and to calculate the properties of the system by some average over all the particles. In front of this impossible task, it was necessary to proceed in another way. One has to leave the microscopic individual picture and to rely on a statistical approach.

The theory is based on a fundamental hypothesis. It is possible to formulate it as follows. The matter, whatever its state, is a compound of microscopic entities which have specific characteristics. The state of each entity does not remain the same but changes with time. In other words, the microscopic world is disordered. However, it is admitted that the collective properties do remain stationary. For example, the molecules of a gas change constantly their velocity because of the collisions between them but their mean velocity is well-defined. In a stationary state, the macroscopic quantities can be found by some averaging procedure.

A particular theory was developed which is based on some postulates. The test of the validity of this method is the comparison between the consequences drawn from the postulates and the experimental results. The very good agreement, which was found, is a guarantee of the validity of the method.

An important consequence of this approach is that thermal phenomena have their origin in mechanics. It is not an obvious thinking and the reader must be ready to adopt it. If a system is made of a relatively small number of bodies, one can apply the law of mechanics like in the planetary system. But if now the system is a compound of a very large number of units, one has thermal phenomena. Some subtle points remain in this approach and they were in the past subject of intense debates. In this book, we shall not consider them except in the historical part. We think that in the first contact with this field, it is better to consider only the basic concepts. We hope that this first encounter with statistical mechanics will help the reader to be able to read more advanced books.

The book is divided into two parts: first the principles of the theory and in a second part some applications. This does not correspond to the historical development of statistical mechanics as it is frequently presented in order to follow the history, we added a chapter which presents the scientists who contributed to the main steps in the development of the theory.

The Thermodynamic Potentials

Before embarking on the exposition of the theory and its applications, important results concerning the thermodynamic potentials are recalled.

If the equilibrium state of the system is defined by the knowledge of some variables, it exits a function of these variables from which all the properties of the system can be deduced. These functions play the role of a potential for the following reasons. If a perturbation appears in the system and the chosen variables are kept constant, the equilibrium state is reached when the potential is a minimum. We shall consider three cases, which are important for the theory of statistical mechanics.

The entropy as a thermodynamic potential

In the first case, the state of the system is controlled by extensive variables like the energy E, the volume V, the number of particles N, etc. (We took only these three variables by convenience). In such case, one considers the system as a closed system since no energy or no particle can enter or leave the system. The extensive variables are those which are proportional to the size of the system. The thermodynamic potential associated with such a state is the entropy $S(E, V, N)$. We recall that in a closed system, equilibrium is reached when S is maximum or when $-S$ is minimum. Entropy is a complex concept which may be presented in several ways. In the framework of this book, entropy is defined as a potential thermodynamic in the particular context of the closed system.

The entropy is a homogeneous function such that

$$S(\lambda E, \lambda V, \lambda N) = \lambda S(E, V, N) \tag{1}$$

where λ is a scalar. From the differential of E

$$dE = TdS - PdV + \mu dN \tag{2}$$

one gets the differential of S

$$dS = dE/T + (P/T)dV - (\mu/T)dN \tag{3}$$

(T is temperature, P the pressure and μ the chemical potential). One sees that the temperature is given by

$$(\partial S/\partial E)_{V,N} = 1/T \tag{4}$$

The function of state $P(V, T, N)$ is deduced from $T(E, V, N)$ and $P(E, V, N) = T(\partial S/\partial V)_{E,N}$ and eliminating E from them.

The Helmotz free energy as a potential

The second case corresponds to the system in thermal contact with a reservoir which defines the temperature of the system with volume V and number of particles N. It is supposed that: (a) The ensemble reservoir + system is a closed system. (b) The reservoir is much larger than the system. The thermodynamic potential is the Helmotz free energy, which is also a minimum when T, V and N are kept constant. Its definition is

$$F(T, V, N) = E - TS \tag{5}$$

and the differential of F is

$$dF = dE - d(TS) = TdS - PdV + \mu dN - TdS - SdT$$

or

$$dF = -SdT - PdV + \mu dN \tag{6}$$

One gets the entropy

$$S(T, V, N) = -(\partial F/\partial T)_{V,N} \tag{7}$$

and the energy

$$E(T, V, N) = F - T(\partial F/\partial T)_{V,N} \tag{8}$$

The function of state is merely $P(T, V, N) = -(\partial F/\partial V)_{T,N}$.

In the case of magnetic material one has to take into account the magnetization of the material under application of a magnetic field. The magnetization M is the number of effective magnetic dipoles and the contribution to the energy is MdH. The differential of E is now

$$dE = TdS - PdV + \mu dN + HdM \tag{9}$$

since E is a function of the extensive quantities as M is.

The free energy is $F_M = E - TS - MH$ such that its differential is

$$dF_M = -SdT - PdV + \mu dN - MdH \tag{10}$$

In other words, F_M is a function of T, V, N and H.

The magnetization is obtained from F_M through its derivative relatively to $H, M = -(\partial F_M/\partial H)_{T,V,N}$.

The grand potential

In the third case, the system is in thermal contact with a reservoir made of the same kind of particles. But now it is enclosed in a cell of volume V with walls permeable to the particles. The variables which characterize the system are the volume V, T (imposed by the reservoir), and the chemical potential which is the same for the system and the reservoir. If one particle crosses the walls from the reservoir to the system when nothing else is changed, the energy of the reservoir decreases by an amount equal to $-\mu_R$ (chemical potential of the reservoir) when the energy of the system increases by μ_s (chemical potential of the system). Since the total energy has not changed (the reservoir + the system are a closed system), it results equality of the chemical potentials.

In the present case the thermodynamic potential is the grand potential

$$\Psi(T, V, \mu) = E - TS - \mu N. \tag{11}$$

From the fundamental equality of thermodynamics, $E = TS - PV + \mu N$ one gets $\Psi = -PV$. The differential of Ψ is $d\Psi = dE - d(TS) - d(\mu N)$ or

$$d\Psi = -SdT - PdV - Nd\mu \tag{12}$$

Thus, one has

$$S = -(\partial\Psi/\partial T)_{V,\mu} \tag{13}$$

$$P = -(\partial\Psi/\partial V)_{T,\mu} \tag{14}$$

and

$$N = -(\partial\Psi/\partial\mu)_{T,V} \tag{15}$$

The energy is given by

$$E = \Psi + TS + \mu N = \Psi - T(\partial\Psi/\partial T)_{V,\mu} - \mu(\partial\Psi/\partial\mu)_{T,V}. \tag{16}$$

The distributions

Thermodynamics look for the relationships between thermal and non-thermal quantities like energy, temperature, volume, pressure, electric and magnetic fields, etc. Each of the three thermodynamics presented above are function of three quantities. In thermodynamics, they are macroscopic quantities that one can measure. They are also well-defined.

One has to make important remarks. The three thermodynamic potentials mentioned above are not the only potentials one can define. There are others like the enthalpy or the free energy of Gibbs, Each potential is considered as a function of some variables (called constrained variables) which give complete differentials from which other variables may be known. All these potentials are equivalent since one can choose the constrained variables in a convenient manner. The constrained variables of a given potential may be deduced variables of another potential.

However, in the frame of statistical mechanics, the results do not give well-defined quantities but only a distribution of the values of different quantities. This is due to the disordered character of the microscopic world. We shall get only probabilities of occurrence of particular values of the parameters.

We shall see three different approaches for passing from the microscopic world to the macroscopic. In the first case, the temperature and the pressure are not controlled by some external influence and one considers the system as a closed system. In the second case, the macroscopic energy is the mean value of the energy of the system. It can fluctuate since only the temperature is imposed by the external reservoir. But it is a fundamental hypothesis that the macroscopic energy that can be measured is the mean value and that the fluctuations around the mean value are negligible when the number of particles is very large. And in the third case, the macroscopic energy is the mean energy and the macroscopic density corresponds to the mean value of N. Since the walls are permeable to the particles, their number in the volume V of the system can vary. If we suppose that the fluctuations are negligible, the three cases give the same results and one can use the standard thermodynamics formulas.

The fundamental goal of statistical mechanics is to determine the thermodynamic potentials S, F and Ψ from the knowledge of the microscopic properties of the particles. It is the subject of the two following chapters.

Chapter 1

The Closed System or the Microcanonical Ensemble

1.1 The Microcanonical Ensemble

In a closed system, N particles are enclosed in a volume V with walls impervious to heat such that the system cannot receive energy from the external world or send energy outside. The energy of the system is E. We suppose also that the system is in equilibrium. In other words, it is supposed that it is possible to prepare the system with well-define values of the energy E, the volume V, the number of particles N and other quantities which we design by α. We call such a situation a *macrostate* of the system.

At a particular time, a particular particle is characterized by a set of variables which define completely its state. For example, the state of a free and isolated particle is given by the knowledge of its momentum vector $\mathbf{p}$. But the energy is not a characteristic of the state since there are several states with the same energy and the same absolute value of the momentum p but with different directions of the vector $\mathbf{p}$. But in a group of several particles, each particle does not stay always in the same state and there is a perpetual change from state to state.

At a particular time, the system is in a state called a *microstate* when the states of all the particles are well defined. We recall the fundamental hypothesis that the particles change their state with time, so the system also passes from one microstate to another but keeps the same macrostate defined by the chosen values of the quantities E, V and N. If we consider an ensemble (called the microcanonical ensemble) of a very large number of identical systems (prepared as above with the same set of variables E, V, N and α), each system is in a particular microstate when all these microstates correspond to the same *macrostate*.

We shall give a simple example of a system of three particles which can have one of four possible energies: 0, ε, 2ε and 3ε. We suppose that the total energy of the system is 3ε and this energy defines the microstates of the system. One can have the following possibilities, each defining a particular microstate: (a) the three particles are in the state with energy ε; (b) one particle is a state with energy 3ε and the two others are in the state with energy 0; (c) one particle is in the state with energy 2ε, one particle in the state with energy ε and the third in the state with energy 0.

The fundamental quantity that we shall define and study is the entropy. In thermodynamics, the concept of entropy has some mysterious aspects. It can be seen as a tool which is very useful but its physical meaning is not very clear. In statistical mechanics, it has a physical meaning and I shall present it as a postulate.

Now, we can state several postulates concerning the ensemble.

Postulate 1: The probability p_i to find a particular system in the ensemble in a given microstate (labeled i) is the same for all the microstates (in number Ω). The probability is a number smaller than 1 such that the sum of all the probabilities is equal to 1. One has $\Sigma p_i = 1$ when the sum has Ω terms. Since p_i is the same for all the microstates, $\Sigma p_i = \Omega p_i = 1$ and one gets $p_i = p = 1/\Omega$.

In other words, there is no preferential microstate for the chosen macrostate. We can already emphasize that the number of microstates Ω is a function of the parameters E, V, N and α.

Postulate 2 of Boltzmann: The entropy S of the system is

$$S = k_B \operatorname{Ln} \Omega \tag{1.1}$$

It is the most fundamental formula of statistical mechanics. Boltzmann first proposed it and it is why the constant k_B is called the Boltzmann constant. Its value is 1.38×10^{-23} Joule $\cdot$ deg^{-1} or 8.62×10^{-5} eV $\cdot$ deg^{-1}. (We recall that 1 eV $= 1.602 \times 10^{-19}$ J). We shall see later how this value was obtained.

The problem is now to show that this postulate gives the same entropy as that defined in thermodynamics. Later, we shall see that we need a new postulate that we formulate in the following.

An important consequence of this postulate 2 is that, for $T = 0$, the entropy is null (except for some particular exceptions). $T = 0$ is the lowest temperature and all the particles are in their lower energy state. So, the number of microstates is 1 since all the particles are in the same state.

Postulate 3: In a closed system, the equilibrium state corresponds to the largest value of the entropy.

In other words, since E, V and N are fixed, the equilibrium state will be given by the value of α which gives the largest value of S.

Now, we shall take two simple examples.

Example 1. One considers a row of six spins aligned. If n_1 spins are up and n_2 spins are down, the number of microstates associated with a particular choice of n_1 and n_2 is

$$\Omega = 6!/(n_1!\ n_2!) \tag{1.2}$$

It is the number of possibilities to put 6 particles in two boxes. If there is no external magnetic field or interaction between spins, the energy is null. The equilibrium state is the one with the largest number of microstates Ω because in this case the entropy is maximum. In the present case, one can choose α as the number n_1 (or n_2). Varying n_1 one sees that Ω is maximum for $n_1 = 3$ ($\Omega = 20$), as shown in Table 1.1.

In the equilibrium state, half of the spins are up and half down. The entropy is equal to $k_B \,\mathrm{Ln}\, 20$.

Example 2. Three spins are located at the three corners of an equilateral triangle. The energy of the systems is only due to the interactions of the spins. m_1, m_2 and m_3 are the values of the three spins

Table 1.1. Number of microstates when the number of spins up is changed.

n_1	0	1	2	3	4	5	6
Ω	1	6	10	20	10	6	1

which can be equal to 1 (spin up) or -1 (spin down). The energy of the system is $E = -(m_1m_2 + m_2m_3 + m_3m_1)$. There are $2 \times 2 \times 2$ microstates depending on the directions of the spins. There are two microstates with energy $E_1 = -3$ (the three spins up or the three spins down) and six microstates (when two spins are in the same direction and the third in the opposite) with energy $E_2 = -1$. In other words, the system can have only two macrostates. What are the possible values of the entropy for such systems? In the first case, the entropy is $S_1 = k_B$ Ln 2, and in the second case, $S_2 = k_B$ Ln 6. Note that the macrostate with the larger energy has also the larger entropy.

1.2 Properties of the Entropy

2.A We consider an isolated system of N particles with energy E in the volume V which is divided into two distinct cells (left cell with volume V_1 and right cell with volume V_2; $V_1 + V_2 = V$). The two cells are separated by a rigid wall covered by a material which does not give the possibility for heat to go from one cell to the other nor for the particles to pass from one side to the other (if it is a gas). On the left side, there are N_1 particles with energy E_1 and N_2 particles with energy E_2 on the right side ($N_1 + N_2 = N$ and $E_1 + E_2 = E$). The two cells are two closed subsystems (Fig. 1.1).

We shall call $\Omega(E, V, N)$ the number of microstates of the whole system, $\Omega_1(E_1, V_1, N_1)$ the number of microstates of the left subsystem and $\Omega_2(E_2, V_2, N_2)$ the number of the microstates of the right subsystem. One has

$$\Omega = \Omega_1\Omega_2 \tag{1.3}$$

since for each microstate of the left subsystem it is possible to associate all the microstates of the right. We conclude from (1.1)

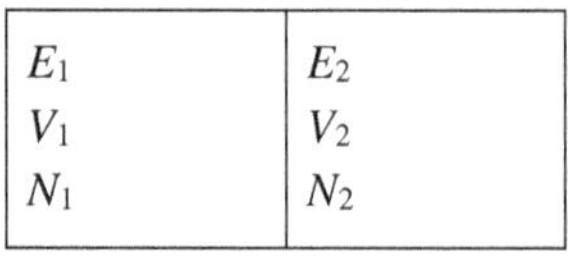

Fig. 1.1. A system and its two subsystems.

that

$$S = k_B \,\mathrm{Ln}\, \Omega = k_B \,\mathrm{Ln}\, (\Omega_1 \Omega_2) = k_B \,\mathrm{Ln}\, \Omega_1 + k_B \,\mathrm{Ln}\, \Omega_2 \tag{1.4}$$

or

$$S = S_1 + S_2 \tag{1.5}$$

The entropy is an additive quantity as it must be.

2.B One considers again the isolated system divided into two subsystems as above. At a given time, one removes the material, which covers the separation wall making it permeable to heat. The two subsystems being different were not at the same temperature, but now that the wall is permeable to heat, there is heat transfer from the side with upper temperature to the side with lower temperature. For $t = \infty$, the two sides reach the same temperature. At $t = 0$ when one removes the isolating material, the system is not in equilibrium and reaches it at the end of the process. At the equilibrium, the entropy reaches its largest value.

To understand this point, one writes (with V and N constant)

$$S = S_1(E_1) + S_2(E_2) = S_1(E_1) + S_2(E - E_1) \tag{1.6}$$

since the total energy E is constant. During the heat transfer process, one can take E_1 as the quantity α which gives S maximum. This condition is

$$(\partial S/\partial E_1)_{V,N} = (\partial S_1(E_1)/\partial E_1)_{V,N} + (\partial S_2(E - E_1)/\partial E_1)_{V,N} = 0 \tag{1.7}$$

or

$$(\partial S/\partial E_1)_{V,N} = (\partial S_1/\partial E_1)_{V,N} + (\partial S_2/\partial E_2)_{V,N}(\mathrm{d}E_2/\mathrm{d}E_1) = 0 \tag{1.8}$$

But $\mathrm{d}E_2/\mathrm{d}E_1 = -1$ since $E_2 = E - E_1$.

The condition $(\partial S/\partial E_1)_{V,N} = 0$ which gives the condition of equilibrium can be written as

$$(\partial S_1/\partial E_1)_{V,N} = (\partial S_2/\partial E_2)_{V,N} \tag{1.9}$$

Since the equilibrium is given by the equality of the temperatures $T_1 = T_2$, one concludes that equality (1.9) is equivalent to the equality of the temperatures. It means that the derivatives $(\partial S_i/\partial E_i)_{V,N}$ are functions of T only. Now, we add the last postulate:

Postulate 4: The derivative of the entropy relative to the energy is equal to the inverse of the temperature

$$(\partial S/\partial E)_{V,N} = 1/T \tag{1.10}$$

Equation (1.10) gives the link with the thermodynamic definition of the entropy. It is not difficult to see that the derivative $(\partial S/\partial E)_{V,N}$ is positive: if one increases the energy of the system, keeping V and N constant, the possible energy for each particle is also increased giving more possible states to it. This means that the total number of microstates has also been increased.

It remains to show that in the process of equality of the temperatures there is an increase of the entropy. It is a well-known result of thermodynamics that an irreversible process takes place with an increase in entropy. For that we compare the entropy S_i at the beginning of the process with the entropy S_f at its end.

At the beginning, the number of microstates is given by (as shown above)

$$\Omega_i = \Omega_1(E_1)\Omega_2(E - E_1) \tag{1.11}$$

However, in the final state, the energy E_1 of the side one is not fixed but can take several values (due to some fluctuations in the temperatures) keeping the sum $E_1 + E_2$ constant. For the calculation of the number of microstates, we have to sum over all the possible values E_i of the side one. We get

$$\Omega_f = \Sigma\Omega(E_1)\Omega(E - E_1) \tag{1.12}$$

In the expression (1.12), there is a term in the sum with the E_i equal to the initial energy E_1 and other terms. This means that the sum (1.12) is larger than (1.11). In other words, the number of microstates in the final state is larger than in the initial state. We conclude that the entropy increased in the process of equalization of temperatures.

2.C One considers the same system as above divided into two subsystems. In the beginning, the wall is clamped such the pressures on both sides are not equal. However, at a given time, the wall is freed and moves until the pressures on both sides are equal. At the same time, the wall is also made permeable to heat so the temperature on both sides will be equal. Taking V_1 and E_1 as quantities for which one looks for equilibrium by maximizing the entropy we write $\partial S/\partial E_1 = 0$ and $\partial S/\partial V_1 = 0$. One gets the following equalities (following the same method as above):

$$(\partial S_1/\partial E_1) = (\partial S_2/\partial E_2) \tag{1.13}$$

and

$$(\partial S_1/\partial V_1)_{E,N} = (\partial S_2/\partial V_2)_{E,N} \tag{1.14}$$

which are equivalent to the equality of the pressures and the temperatures. We have already seen that (1.13) is equivalent to $T_1 = T_2$. To translate the equivalence of (1.14), we need to take into account the dimension of the ratio (entropy/volume). From (1.10), one sees that the dimension of S is (energy/temperature) or (volume $\times$ pressure/temperature). Thus, the dimension of the ratio (entropy/volume) is (pressure/temperature). We can replace the equality (1.14) with the following equality:

$$P_1/T_1 = P_2/T_2 \tag{1.15}$$

The conclusion is that

$$(\partial S/\partial V)_{E,N} = P/T \tag{1.16}$$

2.D Once again, one takes the system divided into two subsystems. Up to now, the physical state of the system (solid, liquid or gas) was not important since the number of particles N_1 and N_2 were constant. Now, we suppose that it is a fluid with the wall not permeable to particles. At a given time, one makes the wall (a) permeable to heat, (b) mobile and (c) permeable to particles. The equilibrium is reached when (a) the temperatures are equal, (b) when the pressures are equal and (c) when the chemical potentials of both sides are equal:

$$\mu_1 = \mu_2 \tag{1.17}$$

To see that, we repeat again the argument presented above. Consider the system at equilibrium when some particles cross the wall (although the mean numbers of particles on both sides are constant, there is always possibility for particles to cross the wall and make the instantaneous numbers different from the mean values). On side 1, the number of particles increases by $\mathrm{d}N_1$ and on the other side by $\mathrm{d}N_2$ with $\mathrm{d}N_1 + \mathrm{d}N_2 = 0$. The changes in the energies of the two sides are respectively $\mathrm{d}E_1 = \mu_1\,\mathrm{d}N_1$ and $\mathrm{d}E_2 = \mu_2\,\mathrm{d}N_2$. Taking into account that the total energy is constant ($\mathrm{d}E_1 + \mathrm{d}E_2 = 0$) and that $\mathrm{d}N_1 = -\mathrm{d}N_2$, one gets equality (1.17).

We write again the condition of equilibrium of maximum of the entropy taking as the variables α, E_1, V_1 and N_1. Proceeding as above, one gets the equalities (1.13), (1.14) and a new equality

$$(\partial S_1/\partial N_1)_{E,V} = (\partial S_2/\partial N_2)_{E,V} \tag{1.18}$$

which is equivalent to (1.17). Considering that the dimension of S is (energy/temperature), that μ is an energy and the fact that N is without dimension, one obtains

$$(\partial S/\partial N)_{E,V} = -\mu/T \tag{1.19}$$

The reason for the negative sign is not trivial and can be justified in the following way using the theorem which gives the relation between the three partial derivatives of three quantities related by a relationship. Let be these three quantities x, y, z and the relation can be written in three equivalent forms: $x(y,z)$, $y(x,z)$ or $z(x,y)$. It is possible to show that (this demonstration is very often given in textbooks of thermodynamics)

$$(\partial x/\partial y)_z(\partial y/\partial z)_x(\partial z/\partial x)_y = -1 \tag{1.20}$$

We shall use this relation for the entropy seen as a function of E and N, keeping V as a constant and we get

$$(\partial S/\partial E)_{N,V}(\partial E/\partial N)_{S,V}(\partial N/\partial S)_{E,V} = -1 \tag{1.21}$$

or introducing $(\partial S/\partial E)_{N.V} = 1/T$ and the definition of $\mu = (\partial E/\partial N)_{S.V}$, this gives

$$(1/T)(\mu)(\partial N/\partial S)_{E,V} = -1 \tag{1.22}$$

from which one deduces (1.19). Without negative sign, it should not be possible to respect theorem (1.20).

From expressions (1.10), (1.16) and (1.19), we get the differential of the entropy

$$\mathrm{d}S = \mathrm{d}E/T + (P/T)\mathrm{d}V - (\mu/T)\mathrm{d}N \tag{1.23}$$

1.3 An Example

We now solve a particular situation of N particles in a solid with two possible states when one supposes that N is a very large number. Each state is characterized by its energy: state 1 with energy 0 and state 2 with energy e. In the present case, the volume is constant and does not play any role, so we drop it. And the number of particles being constant, the different interesting quantities are functions of T only. If one considers the system as a closed one, it is possible to apply the method of the microcanonical ensemble. In a given macrostate, the number of particles in each state is specified, and this corresponds to a particular value of the energy and also to the temperature. Taking a different macrostate (i.e. a different distribution of the particles between the two possible states), one gets a different value of the energy and of the temperature. The question is to find the energy, the entropy and the number of particles in each particle state as functions of the temperature. First, one calculates the entropy.

If n particles are in state 2, the energy of the system is $E = ne$. The number of microstates is given by

$$\Omega = N!/[n!(N-n)!] \tag{1.24}$$

It is the number of manners to choose n particles among N particles. It is equal to the number of manners to put N particles in two boxes, n particles in the first and $N-n$ particles in the second. And the entropy is

$$S = k_B \,\mathrm{Ln}\, \Omega = k_B[\mathrm{Ln}\, N! - \mathrm{Ln}\, n! - \mathrm{Ln}(N-n)!] \tag{1.25}$$

To go further, we adopt the Stirling approximation: $\mathrm{Ln}\, N! \approx N\,\mathrm{Ln}\, N - N$, which is good, even for a relatively small number, say 10^4. In the present case of a macroscopic body, n and N may be very large (order of 10^{19}) and, as we see in the following, even when the used approximation is not valid, the results are qualitatively correct.

From (1.25), one gets

$$S = k_B[N \ln N - n\,\mathrm{Ln}\, n - (N-n)\,\mathrm{Ln}\,(N-n)] \tag{1.26}$$

In order to find the functions $E(T)$, $S(T)$ and $n(T)$, one writes the relation $(\partial S/\partial E) = 1/T$ using $\mathrm{d}S = (\partial S/\partial n)_E\, \mathrm{d}n$ and $\mathrm{d}E = e\,\mathrm{d}n$.

This gives (writing $\mathrm{d}n = \mathrm{d}E/e$)

$$(\partial S/\partial E)_n = (\partial S/\partial n)_E(\partial n/\partial E) \tag{1.27}$$

or

$$(\partial S/\partial E)_n = (1/e)(\partial S/\partial n)_E = 1/T \tag{1.28}$$

From (1.26), one calculates the derivative $(\partial S/\partial n)_E$ and the result following (1.28) is $(\partial S/\partial n)_E = e/T$

$$(\partial S/\partial n)_E = k_B \,\mathrm{Ln}[(N-n)/n) = e/T \tag{1.29}$$

From (1.29), one can extract the relation between n and T

$$n = N/[1+\exp(e/k_B/T)] \tag{1.30}$$

Now, it is easy to insert $n(T)$ in the expressions of the energy and of the entropy. One has for E,

$$E = ne = (Ne)/[1+\exp(e/k_b T)] \tag{1.31}$$

and for S,

$$\begin{aligned} S = k_B N[&(1+\exp(e/k_B T))^{-1}\,\mathrm{Ln}(1+\exp(e/k_B T)) \\ &+(1+\exp(-e/k_B T))^{-1}\,\mathrm{Ln}(1+\exp(-e/k_B T))] \end{aligned} \tag{1.32a}$$

which can be also written in the following form:

$$S = k_B N(e/k_B T)[1+\exp(e/k_B T)]^{-1} + k_B N\,\mathrm{Ln}[1+\exp(-e/k_B T)] \tag{1.32b}$$

The two expressions (1.32a) and (1.32b) look so different that one can doubt if it is really the same function. It is left as an exercise for the reader to find a way to pass from one expression to another. We suggest to use the following identities: $(1+1/x)^{-1} + (1+x)^{-1} = 1$ and $\mathrm{Ln}[(1+e^x)/(1+e^{-x})] = x$. The complete derivation is given at the end of the chapter.

Now, it is interesting to look for the limits of n and S at low and high temperatures. The low temperatures are defined by the condition $k_B T \ll e$ or $(e/k_B T) \gg 1$. This gives for n (since $e/k_b T$ goes to infinity and $\exp(e/k_b T) \to \infty$)

$$n = N \exp(-e/k_B/T) \tag{1.33}$$

and one sees that n goes to zero as T goes to zero.[1] This result was expected since for $T = 0$ all the particles are in the state one with the lowest energy and n goes to 0. So, the total energy is also zero.

For the entropy, one has to take only the first term of (1.32b) (the second term goes to 0 (because $\exp(-e/k_B T)$ goes to 0) and this first term becomes

$$S = k_B N (e/k_B T) \exp(-e/k_B T) \tag{1.34}$$

which goes to zero for $T = 0$ (recalling that the product $x \exp(-x)$ goes to zero if x goes to ∞) as expected since there is only one microstate.

Now, at high temperatures, when $k_B T \gg e$, one obtains (since $e/k_b T \ll 1$, one can use the approximations valid for $x \ll 1$: $\exp(x) \approx 1 + x$ and $1/(1+x) \approx 1 - x$)

$$n = N/(2 + e/k_B T) = (N/2)[1 - e/(2k_B T)] \tag{1.35}$$

and

$$S = N k_B [\text{Ln } 2 + (e/2k_B T)^2] \tag{1.36}$$

As T goes to infinity, n tends to $N/2$, and there is equal distribution of the particles between the two states and S tends to $N k_B \text{ Ln } 2$.

In Fig. 1.2, we give the variations of (n/N) and $(S/k_B N)$ with T when one chooses $e/k_B = 10$.

Important Remark: In the determination of the number of microstates, one has admitted that it is possible to follow separately each particle. Consequently, when there are n particles in level 2, if one permutes one particle from level 1 with one particle from level 2, one has the same energy but a different microstate. We say that the particles are distinguishable. We did not introduce the mechanical

[1] In such a case, the Stirling approximation is not always applicable. Nevertheless, the result is qualitatively correct.

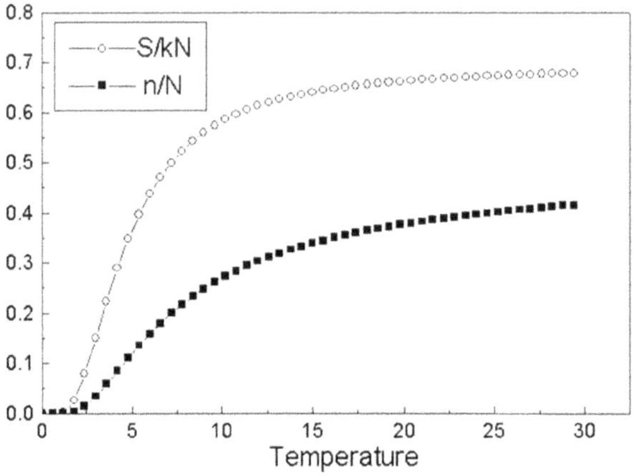

Fig. 1.2. Variations of the entropy and the relative number of particles in the state with the largest energy as functions of the temperature. The ratio n/N tends to 0.5 when the ratio (S/k_BN) tends to Ln $2 \sim 0.7$.

quantum behavior of the particles, thus in this example, the particles are classical and can be distinguished.

Equivalence of (1.32a) and (1.32b)

Writing $x = e/k_BT$, (1.32) can be written as

$$S/k_BN = \text{Ln}(1+e^x)/(1+e^x) + \text{Ln}(1+e^{-x})/(1+e^{-x}) \qquad (1.37)$$

Using the identity $\text{Ln}[(1+e^x)/(1+e^{-x})] = x$ one can write

$$\text{Ln}(1+e^x) - \text{Ln}(1+e^{-x}) = x$$

or

$$\text{Ln}(1+e^x) = x + \text{Ln}(1+e^{-x}) \qquad (1.38)$$

One introduces this expression of $\text{Ln}(1+e^x)$ in (1.37) and one gets

$$S/k_BN = [x + \text{Ln}(1+e^{-x})]/(1+e^x) + \text{Ln}(1+e^{-x})/(1+e^{-x}) \qquad (1.39)$$

One sees that one can develop (1.39) and regroup the terms as

$$S/k_BN = x/(1+e^x) + \mathrm{Ln}(1+e^{-x})[1/(1+e^x) + 1/(1+e^{-x})] \tag{1.40}$$

Now, one uses the second identity $1/(1+e^x) + 1/(1+e^{-x}) = 1$ and one gets

$$S/k_BN = x/(1+e^x) + \mathrm{Ln}(1+e^{-x}) \tag{1.41}$$

which is (1.32b) when x is replaced by e/k_BT.

Chapter 2

The System in Thermal Contact with a Reservoir (the Canonical Ensemble and the Grand Canonical)

In this chapter, we shall see the most important concept of statistical mechanics, the partition function. This method is much often used than that presented in the microcanonical ensemble.

This chapter is also relatively difficult and we suggest to the reader to read it step by step without skipping some lines.

There is a novelty concerning the concept of chemical potential. In thermodynamics it appears when several phases are in contact, i.e. it is related to more than one phase. In statistical mechanics it receives a new role even for an isolated ubique system.

We begin by a system in contact with a very large system at constant temperature T. We shall call it the reservoir. In a first case, the contact is purely thermal; it means that the temperature of the system is fixed by the contact with the reservoir. In the second case, the contact is thermal but there is also the possibility of exchanging particles between the system and the reservoir. Since it is supposed that the system and the reservoir are in equilibrium, there are equalities of the temperatures and of the chemical potentials, as explained above. Our goal is always to find the relation between the microscopic properties of the particles of the system and its macroscopic properties.

The fundamental hypotheses are:

(a) The system and the reservoir are seen together as a grand system which is isolated from the rest of the world. The results of the preceding chapter can be applied to the grand system. In particular, its energy E_0 is constant and the entropy is given by (1.1).

(b) The system is much smaller than the reservoir itself. In particular the energy E_S of the system is much smaller than the energy E_R of the reservoir and that of grand system. Thus one will be able to apply approximations concerning small quantities versus large ones.

2.1 The Canonical Ensemble

2.1.1 *The partition function*

As exposed above, the system with volume V and number of particles N is in thermal contact with the reservoir, which imposes the temperature. However, the energy of the system is not fixed by the contact with the reservoir and can fluctuate over all the possible energies of the system. Our first goal will be to determinate, in the ensemble of a very large number identical systems which constitutes the canonical ensemble, the probability to find one system in a microstate labeled s with energy E_S. Once this probability will be found, we shall be able to calculate the mean energy, which is the macroscopic energy of the system and other quantities.

In the case of the thermal contact of the system with the reservoir, it was stated that the temperature is fixed when the energy is fluctuating. This means that there is constant transfer of energy from the reservoir to the system and vice versa. How is this possible if there is no temperature difference between the system and the reservoir? One possibility is the transfer of potential energy through the variations of the distances between particles since this kind of energy is dependent on distances. But in fact there are changes in the temperatures although they are so small that one does not consider them and accepts to say that the temperature is constant.

If one picks at random a system in the ensemble, the probability p_s to find this system in a microstate s with energy E_S is the ratio of two quantities: first the number of microstates of the system with energy E_S and secondly the total number of microstates of the grand system. A probability is defined as the number of “favorable” cases (here a favorable case is the system is a microstate with energy E_S) divided by the number of all the possible cases (here number of microstates of the grand ensemble, favorable or not).

If the energy of the system is E_S and that of the grand system is E_0, the energy of the reservoir is $(E_0 - E_S)$ and the number of microstates in which the system has energy E_S is equal to the number of microstates in which the reservoir has the energy $E_0 - E_S$. One writes

$$p_s = \Omega_R(E_0 - E_S)/\Omega_{GS}(E_0) \tag{2.1}$$

Ω_R indicates the number of microstates of the reservoir and Ω_{GS} those of the grand system. From the preceding chapter, we know that the entropies S_{GS} of the grand system and that S_R of the reservoir are

$$S_{GS} = kB \text{ Ln } \Omega_{GS}(E_0)$$
$$S_R = kB \text{ Ln } \Omega_R(E_0 - E_S)$$

or

$$\Omega_{GS}(E_0) = \exp[S_{GS}(E_0)/kB] \tag{2.2}$$
$$\Omega_R(E_0 - E_s) = \exp[S_R(E_0 - E_S)/kB] \tag{2.3}$$

For the grand system there is no problem to apply the preceding formula since it is a closed system. For the reservoir, it is only an approximation since it is in contact with the system. It is a good approximation since it is much larger than the system.

From (2.1), (2.2) and (2.3) one gets

$$p_s = \exp[S_R(E_0 - E_S)/kB]/\exp[S_{GS}(E_0)/kB] \tag{2.4}$$

One can expand $S_R(E_0 - E_S)$ as a Taylor development[1] and stop it after the second term since $E_0 \gg E_S$

$$S_R(E_0 - E_S) = S_R(E_0) - E_S(\partial S_R/\partial E)_0 \tag{2.5}$$

[1]The Taylor development of a function $f(x)$ is given by $f(x+h) = f(x) + h\, df/dx(x) + (h^2/2!)d^2f/dx^2(x) + \cdots + (h^n/n!)d^n f/dx^n(x) +$

But $(\partial S_R/\partial E) = 1/T$

$$S_R(E_0 - E_s) = S_R(E_0) - E_S/T \tag{2.6}$$

The probability p_s can be written as

$$p_s = \exp[S_R(E_0)/k_B - E_S/(kBT)]/\exp[S_{GS}(E_0)/kB]$$

or

$$p_s = \exp[S_R(E_0)/kB]\exp[-E_S/(kBT)]/\exp[S_{GS}(E_0)/kB] \tag{2.7}$$

Since the sum of all the probabilities is equal to 1, this means that

$$\Sigma_s p_s = \Sigma_s \exp[S_R(E_0)/kB]\exp[-E_S/(kBT)]/\exp[S_{GS}(E_0)/kB] = 1 \tag{2.8}$$

or

$$\Sigma_s p_s = \{\exp[S_R(E_0)/kB]/\exp[S_{GS}(E_0)/kB]\} \times \Sigma_s \exp[-E_S/(kBT)] = 1$$

Thus

$$\exp[S_{GS}(E_0)/kB] = \exp[S_R(E_0)/kB]\Sigma_s \exp[-E_S/(kBT)] \tag{2.9}$$

Putting (2.9) in (2.7), we get the final result for p_s:

$$p_s = \exp(-E_S/kBT)/\Sigma_s \exp[-E_S/(kBT)] \tag{2.10}$$

when the sums in (2.8), (2.9) and (2.10) are over all the possible microstates and not only on the possible values of the energies. The probability p_s is the probability to find the system in a microstate or in a state with energy E_S. It is an important result and we shall use it several times.

The sum

$$Z = \Sigma_s \exp[-E_S/kBT] \tag{2.11}$$

is called the partition function. In the context of the canonical ensemble, this is the most important result. We shall see that it will give the link between the microscopic and the macroscopic points of view. Z is a function of the parameters that define the system: T, V and N.

The temperature T appears explicitly, but the possible energies of the system depend in general on V and N.

It is usual to take the Greek letter β as the inverse of the temperature $\beta = 1/(k_B T)$ and to write Z as

$$Z = \Sigma_s \exp(-\beta E_S) \tag{2.12a}$$

It is possible that several different microstates have the same energy, thus there is a function $g(E_S)$ giving the number of microstates for a given energy E_S. Such states with the same energy are called degenerate states. The partition function can be written as a sum over the energies:

$$Z = \Sigma_E g(E_S) \exp(-\beta E_S) \tag{2.12b}$$

We recall and stress the difference between (2.11) or (2.12a), and (2.12b). In (2.11) the sum is over all the microstates and in (2.13) it is over the energies.

2.1.2 *The energy, the entropy, the thermodynamic potential*

2.2A With the knowledge of p_s, it is possible to calculate the mean energy E as

$$E = \Sigma_s p_s E_S \tag{2.13a}$$

or

$$E = \Sigma_s E_S \exp(-\beta E_S)/\Sigma_s \exp(-\beta E_S) \tag{2.13b}$$

The formula (2.13a) is the standard formula for the mean value of a quantity which has different probability to appear.

The numerator of (2.13b) is minus the derivative of the denominator relatively to β and the energy can written using Z as

$$E = -[\partial Z/\partial \beta]/Z = -[\partial \mathrm{Ln}\ Z/\partial \beta] \tag{2.14}$$

2.2B The energy E is given as a function of T, V and N. The derivative $(\partial E/\partial T)_{V,N}$ is the specific heat at constant volume, C_V.

The entropy is given by[2]

$$S = \int C_V(dT/T) = \int (\partial E/\partial T)_V (dT/T).$$

One uses the following relations linking T and β to get:

$$T = 1/(kB\beta)$$

$$dT = -d\beta/(kB\beta^2)$$

$$(\partial E/\partial T) = (\partial E/\partial \beta)(d\beta/dT) = (\partial E/\partial \beta)(-kB\beta^2)$$

giving

$$S = \int (\partial E/\partial T)_V (dT/T) kB = kB \int \beta(\partial E/\partial \beta)_V d\beta.$$

Performing integration by parts, one writes

$$u = \beta \quad dv = (\partial E/\partial \beta)_V d\beta \int u\ dv = uv - \int v\ du$$

$$du = d\beta \quad v = E$$

One gets, using (2.14)

$$S = kB[\beta E - \int E d\beta] = kB[\beta E + \int (\partial \text{Ln } Z/\partial \beta) d\beta] \quad (2.15)$$

or

$$S = E/T + kB\text{Ln Z} \quad (2.16)$$

From (2.16) we have the link that we are looking for

$$E - TS = F = -kBT \text{ Ln Z} \quad (2.17)$$

We recall that F is the Helmotz free energy and it is a function of T, V and N. We mentioned above that from the knowledge of F we

[2]The fundamental relation between the entropy dS and the heat dQ in a infinitesimal process is $dS = dQ/T$. If the process takes place at constant volume $dQ = C_V dT$ and $dS = (C_V/T)dT$.

can get all the possible data about the system. The relation (2.17) is one of the most important of this course.

We close this section by another formulation of the entropy:

$$S = -kB\Sigma_s p_s \mathrm{Ln}\, p_s \tag{2.18a}$$

To show that, one introduces in (2.17) the expression (2.10) for p_s and one gets:

$$S = -kB\Sigma_s[\exp(-\beta E_S)/Z][-\beta E_S - \mathrm{Ln}\ Z] \tag{2.18b}$$

Or

$$S = kB\Sigma_s[\beta E_S \exp(-\beta E_S)]/Z + kB[(\mathrm{Ln}\ Z)/Z]\Sigma_s \exp(-\beta E_S) \tag{2.18c}$$

Taking account of (2.13), one sees that the first term in (2.18b) is equal to E/T. Since $\Sigma_s \exp(-\beta E_S) = Z$ the second term becomes equal to $k_B\mathrm{Ln}\ Z$. Finally (2.18c) is equal to

$$S = E/T + kB\mathrm{Ln}\ Z.$$

One recovers the expression (2.16) for the entropy.

The expression (2.17) is very general and can be used for the closed system of the microcanonical ensemble. In this case, the probability to find the system in a microstate with the chosen energy is $1/\Omega$, since all the microstates, in number Ω, have the same probability to be found in the ensemble. Putting $p_s = 1/\Omega$ in (2.17) gives again the Boltzmann formula (1.1), $S = k_B\mathrm{Ln}\ \Omega$ since the number of terms in the sum is merely Ω.

2.2C We consider N particles without interaction in a volume V at temperature T. We add also that it is possible to distinguish between the particles. This means that it is always possible (in principle) to follow an individual particle. In such case, the permutation of two particles between their respective states introduces a new microstate for the system. This cannot be true for gas in which there are constant collisions between the particles such that their "individuality" is lost.

In the expression of Z will appear all the energies of the system. These energies are all the possible sums of the individual energies (of the microstates)[3] of the N particles. We note by $\{e_i\}$one of these

[3]One can sort the energies of the microstates of one particle as e_1, e_2, $e_3, \ldots$ The energies e_i are not necessarily all different since it is possible that some microstates have the same energy.

possible sums (with N terms) and the partition function is $Z = \Sigma \exp(-\beta\{e_i\})$ over all these $\{e_i\}$

The one particle partition function is $Z_1 = \Sigma \exp(-\beta e_i\}$ where the sum is on the .possible energies of one particle and now consider the following expression B:

$$B = (Z_1)^N = [\exp(-\beta\Sigma_i e_i)]^N \tag{2.19}$$

One can write

$$\begin{aligned} B &= [\exp(-\beta\Sigma_i e_i)]^N \\ &= [\exp(-\beta\Sigma_i e_i)][\exp(-\beta\Sigma_i e_i)][\exp(-\beta\Sigma_i e_i)] \ldots. \end{aligned} \tag{2.20a}$$

when the right-hand side contains N identical terms. More explicitly (2.20a) is

$$\begin{aligned} &[(\exp(-\beta e_1) + \exp(-\beta e_2) + \exp(-\beta e_3) \ldots] \\ &[(\exp(-\beta e_1) + \exp(-\beta e_2) + \exp(-\beta e_3) \ldots] \\ &[(\exp(-\beta e_1) + \exp(-\beta e_2) + \exp(-\beta e_3) \ldots][\ldots. \end{aligned} \tag{2.20b}$$

If one expands these products, one finds that B will be given by a sum of terms of the form $\exp[-\beta\{e_i\}]$. As above $\{e_i\}$ mentions one of the possible sums of the energies of the particles. This gives

$$B = \Sigma \exp[-\beta\{e_i\})$$

and one sees that B is the partition function of the system. Consequently

$$Z = (Z_1)^N \tag{2.21}$$

We shall take a very simple example of a system with 2 particles, each with two energies e_1 and e_2. The possible energies of the system or the different $\{e_i\}$ are: $e_1 + e_1$, $e_1 + e_2$, $e_2 + e_1$ and $e_2 + e_2$. One has

$$\begin{aligned} Z = \exp[-\beta(e_1 + e_1)] + \exp[-\beta(e_1 + e_2)] + \exp[-\beta(e_2 + e_1)] \\ + \exp[-\beta(e_2 + e_2)] \end{aligned} \tag{2.22}$$

or noting that $\exp[-\beta(e_1 + e_1)] = \exp(-\beta e_1)\exp(-\beta e_1)$ one has

$$(a_i = \exp(-\beta e_i)), \quad Z = a_1 a_1 + 2a_1 a_2 + a_2 a_2 = (a_1 + a_2)^2 = (Z_1)^2.$$

Example 1 (A two-level system). We take the same example that we solved at the end of the preceding chapter. We recall that there are, for each particle, two possible states, one with energy 0 and the other with energy e.

We determine the partition function in two ways. The first is to calculate Z for the system of N particles. The second is to use (2.21) and to calculate the one particle partition function.

To be able to compute the partition function, we need to know the number of microstates with a given energy $E = ne$. This number was already calculated and it is equal to $C_{nN} = N!/[(N-n)!n!]$. Thus Z is (using (2.13))

$$Z = \Sigma_n C_{nN} \exp(-\beta ne) = \Sigma_n x^n N!/[(N-n)!n!] \qquad (2.23)$$

with

$$x = \exp(-\beta ne)$$

The sum in (2.21) is the development of the quantity $(1+x)^N$. Thus the final result is

$$Z = [1 + \exp(-\beta e)]^N \qquad (2.24)$$

This result agrees with what we said about the partition function of independent particles, $Z = Z_1^N$. The one-particle partition function is

$$Z_1 = 1 + \exp(-\beta e)$$

since for one particle there are only two states.

The free energy F is

$$F = -k_B T \text{Ln } Z = -k_B TN \text{ Ln}[1 + \exp(-\beta e)].$$

Now it is straightforward to get the energy and the entropy using the formula (2.14) and (2.16) or from the derivatives of F:

$$E = F - T(\partial F/\partial T)_N \quad \text{and} \quad S = -(\partial F/\partial T)_N.$$

One gets

$$E = Ne/[1 + \exp(\beta e)]$$

and

$$S = kB\ N[(e/kBT)/(1 + \exp(e/kBT) + \text{Ln}(1 + \exp(-e/kBT))]$$

as found above (Eqs. (1.31) and (1.32b)).

Finally one calculates the chemical potential $\mu = (\partial F/\partial N)_{T,V}$:

$$\mu = \partial\{-kBTN\mathrm{Ln}[1+\exp(-\beta e)]\}/\partial N = -kBT\mathrm{Ln}[1+\exp(-\beta e)] \tag{2.25}$$

Example 2 (The ideal gas: The equipartition of the energy in classical mechanics). In this section, we consider particles in classical mechanics. The energy of one particle in a volume V is the sum of two terms, the kinetic energy E_K and the potential energy V_p when E_K is a quadratic function of the particle velocities or the linear momentums and V_P is function of the particle positions. But we have a new situation. Until now, the quantities in the calculations have discontinuous values: for example the energy and the entropy. All the results were presented as series. But in the present cases, the energy and other variables have continuous values. We shall adopt the current believing that it is possible to transform directly a series by a definite integral. We shall discuss this point at the end of Chapter 4.

In one dimension (particles on a line of length L) the kinetic energy of one particle is $E_K = p_x^2/2m$ and the partition function is

$$Z = (1/Q)\int\int \exp[-\beta(E_K + V_p)]\,dp_x\,dx \tag{2.26}$$

where the limits of the integrals for the two variables (p_x, x) are: $-\infty$ and ∞ for the momentum and 0 and L for the position. It is the application of the general expression of the partition function

$$Z = \Sigma_s \exp[-E_S/kBT] \tag{2.11}$$

to the case of one particle in classical mechanics. The introduction of the quantity Q is needed since by definition Z is a quantity without dimension but the double integral in (2.26) has the dimension of a (momentum)(length). This is the reason why one has to introduce the quantity Q with this dimension. In fact the dimension of the product (momentum)(length) is that of the product (mass)(velocity)(length) or $[M][v][L]$. This product has the dimension of [energy][time]=$[E][T]$. One can see the following transformations:

$$[M][v][L] = [M[v][L][T]/[T] = [M][v]^2[T] = [E][T]$$

We shall see below what constant was chosen for Q.

This expression can be transformed into the product of two integrals

$$Z = (1/Q) \int_{-\infty}^{\infty} \exp[-\beta(p_x^2/2m)]\,dp_x \int_0^L \exp[-\beta V_p(x)]dx \quad (2.27)$$

Now we consider the case of particles without potential energy and only with kinetic energy. It is the case of the monatomic ideal gas where the atoms have only kinetic energy and there is no interaction between them. This absence of interaction is characteristic of the concept of ideal gas. The one-particle partition function becomes

$$\begin{aligned} Z_1 &= (1/Q) \int_{-\infty}^{\infty} \exp[-\beta(p_x^2/2m)]\,dp_x \int_0^L dx \\ &= (L/Q) \int_{-\infty}^{\infty} \exp\{-[(p_x^2/(2mkBT)]\}\,dp_x \end{aligned}$$

To calculate Z_1 we use the following trick: we divide the integral by $(2mk_BT)^{0.5}$ and multiply it by the same factor. This gives

$$\begin{aligned} Z_1 &= (L/Q)(2mkBT)^{0.5} \\ &\times \int_{-\infty}^{\infty} \exp\{-[(p_x^2/(2mkBT)]\}d[p_x/(2mkBT)^{0.5}] \quad (2.28) \end{aligned}$$

The integral is now a definite integral of the variable $x = p_x/(2mk_BT)^{0.5}$ with the same limits ($-\infty$ and ∞)

$$Z_1 = (L/Q)(2mkBT)^{0.5} \int_{-\infty}^{\infty} \exp(-x^2)dx$$

If there is no potential energy for the particle the partition function becomes

$$Z_1 = (L/Q)A(2mkBT)^{0.5} \quad (2.29)$$

where

$$A = \int_{-\infty}^{\infty} \exp(-x^2)dx = \sqrt{\pi}.$$

The energy is calculated through the formulas

$$E = F - T(\partial F/\partial T) \quad \text{and} \quad F = -kBT\text{Ln}\, Z._1$$

One gets (note that we do not need to need the value of A to calculate the energy)

$$E = (1/2)kBT \tag{2.30}$$

This energy is the mean kinetic energy of the particle and is called the thermal energy of the particle. It is a remarkable result that this means kinetic energy is proportional to the temperature. For N particles one multiplies the above result by N.

In the case of a particle in three dimensions the kinetic energy is $E_K = (p_x^2 + p_y^2 + p_z^2)/2m$. The partition function is now a multiple integral

$$Z_1 = (1/Q)^3 \int \exp\{-\beta[(p_x^2 + p_y^2 + p_z^2)/2m + V_p(x, y, z)]\} dp_x dp_y dp_z dxdydz \tag{2.31}$$

If there is no potential energy it reduces to

$$Z_1 = (1/Q^3) \int \exp\{-\beta[(p_x^2 + p_y^2 + p_z^2)/2m]\} dp_x dp_y dp_z dxdydz \tag{2.32}$$

This can written as the product of three integrals (when the integral on the variables x, y, z is equal to V)

$$Z_1 = (V/Q^3) \int_{-\infty}^{\infty} \exp[-\beta(p_x^2/2m)] dp_x \int_{-\infty}^{\infty} \exp[-\beta(p_y^2/2m)] dp_y \times \int_{-\infty}^{\infty} \exp[-\beta(p_z^2/2m)] dp_z \tag{2.33}$$

And from the above results we get

$$Z_1 = (V/Q^3)(2\pi\mathrm{m}\ kBT)^{3/2} \tag{2.34}$$

and finally

$$E = 3/2\ kBT \tag{2.35}$$

The preceding results can be generalized in the following formulation. If in the total energy of a system there is a term which is quadratic in some parameter like the momentum or the position, this term contributes to the energy by the amount $(1/2)k_BT$. The parameter is called a degree of freedom.

A simple application of this theorem is the one-dimensional harmonic oscillator. Its energy is $E = p^2/2\text{m}+Kx^2/2$. In thermal contact with a reservoir at temperature T, its thermal energy is k_BT since there are two degrees of freedom.

Now we come back to the value of Q. The expression (2.34) is the one particle partition function of the ideal gas (in three dimensions). Below in Chapter 4, we shall calculate this partition function from the quantum mechanics principles. In order to make the two results identical (in this chapter and in Chapter 4) one has to chose $Q = h$, the Planck constant, effectively the Planck constant has the dimension of (energy)(time).

Important remark. The partition function is defined as a series $Z = \Sigma_s \exp[-E_S/k_BT]$ but for its calculation, we transform this expression into an integral:

$$Z = (1/Q) \iint \exp[-\beta(E_K + V_p)]\, dp_x\, dx$$

and we did this without any justification. It is very frequent to find this equivalence series-integral in spite of their differences and we did as it is usual. In fact, the relation between the sum of a series and the sum of the corresponding integral is an old problem in mathematics. It received several solutions and we postponed to Chapter 4 a detailed discussion of this point. It is very likely that this procedure of replacement is due to the fact that it is easier to calculate a definite integral than to calculate the sum of a series.

2.2 The Grand Canonical Ensemble

2.2.1 *The grand partition function*

As above, we consider a large ensemble of identical systems We are looking for the probability p_s to find the system in a microstate s with energy E_S and number of particles n_s. We begin by the same expressions as above in writing p_s as (2.1) but in specifying that the number of microstates are function of the energy and the number of particles:

$$p_s = \Omega_R(E_0 - E_S, N_0 - n_s)/\Omega_{GS}(E_0, N_0) \tag{2.36}$$

N_0 is the number of particles in the grand system and it is a constant and $N_0 - n_s$ is the number of particles in the reservoir. We write again as above

$$S_{GS} = kB\text{Ln}\, \Omega_{GS}(E_0, N_0)$$
$$S_R = kB\text{Ln}\, \Omega_R(E_0 - E_S, N_0 - n_s)$$

or

$$\Omega_{GS}(E_0) = \exp[S_{GS}(E_0, N_0)/kB] \tag{2.37}$$
$$\Omega_R(E_0 - E_S) = \exp[S_R(E_0 - E_s, N_0 - n_s)/kB] \tag{2.38}$$

In the following steps, we use the thermodynamic relation $E = TS - PV + \mu N$ to write S in the following form:

$$S = (PV)/T + (E - \mu N)/T \tag{2.39}$$

and putting it in (2.37) and (2.38) one gets

$$\begin{aligned}\Omega_{GS}(E_0, N_o) &= \exp[(P_{GS}V_{GS})/T + (E_0 - \mu N_0)/kBT] \\ &= \exp[(P_{GS}V_{GS})/kBT]\exp[(E_0 - \mu N_0)/kBT] \\ &= A_{GS}\exp[(E_0 - \mu N_0)/kBT] \end{aligned} \tag{2.40}$$

with

$$A_{GS} = \exp[(P_{GS}V_{GS})/kBT].$$

For $\Omega_R(E_0 - E_s, N_0 - n_s)$ one has

$$\begin{aligned}&\Omega_R(E_0 - E_S, N - n_s) \\ &\quad = \exp[(P_RV_R)/kBT]\exp[(E_0 - E_S)/kBT - \mu(N_0 - n_s)/kBT] \\ &\quad = A_R\exp[(E_0 - E_S)/kBT - \mu(N_{\grave{a}-}n_s)/kBT] \\ &\quad = A_R\exp[(E_0 - \mu N_0)/kBT]\exp[-(\mathrm{E_S} - \mu n_s)/kBT] \end{aligned} \tag{2.41}$$

with $A_R = \exp[(P_RV_R)/k_BT]$.

Putting (2.40) and (2.41) in (2.36) gives

$$p_s = (A_R/A_{GS})\exp[-(E_S - \mu n_s)/kBT] \tag{2.42}$$

The sum of the probabilities over all the microstates $\Sigma_s p_s = 1$, thus

$$\Sigma_s p_s = (A_R/A_{GS})\Sigma_s \exp[-(E_s - \mu n_s)/kBT] = 1$$

and this means that

$$A_{GS}/A_R = \Sigma_s \exp[-(E_s - \mu n_s)/kBT]$$

Finally we get the final result for p_s

$$p_s = \exp[-(E_s - \mu n_s)/kBT]/\Sigma_s \exp[-(E_s - \mu n_s)/kBT] \tag{2.43}$$

And we stress again that the sum is over all the possible microstates of the system with all the possible energies and number of particles. The values of the mean energy and the mean number of particles are

$$E = \Sigma_s p_s E_S \text{ and } N = \Sigma_s p_s n_s$$

But we can get these values through the grand partition function Z_G defined by

$$Z_G = \Sigma_s \exp[-\beta(E_S - \mu n_s)] \tag{2.44}$$

with $\beta = 1/k_B T$. The sum (2.44) is in fact a double sum: on the microstates and on the number of particles. Z_G is a function of T, V and μ.

2.2.2 *The number of particles, the energy, the entropy and the grand potential*

The mean number of the particles N in the system is given by

$$N = \Sigma_s n_s p_s = \Sigma_s n_s \exp[-\beta(E_S - \mu n_s)]/\Sigma_s \exp[-\beta(E_s - \mu n_s)] \tag{2.45}$$

Recalling that the numerator of (2.45) is the derivative of the denominator relatively to μ divided by β and with a change in the sign, (2.45) can be written as

$$N = (1/\beta)(\partial Z_G/\partial\mu)/Z_G = (1/\beta)(\partial \text{Ln } Z_G/\partial\mu) \qquad (2.46)$$

The energy $E = \Sigma_s p_s E_s$ can be calculated with the help of the derivative $(\partial\text{Ln } Z_G/\partial\beta)$ which is

$$(\partial\text{Ln } Z_G/\partial\beta) = -\Sigma_s(E_s - \mu n_s)\exp[-\beta(E_S - \mu n_s)]/ \\ \times \Sigma_s \exp[-\beta(E_s - \mu n_s)] \qquad (2.47)$$

The numerator of (2.47) is

$$-\Sigma_s E_s \exp[-\beta(E_s - \mu n_s)] + \Sigma_s \mu n_s \exp[-\beta(E_s - \mu n_s)]$$

and consequently $\partial\text{Ln } Z_G/\partial\beta)$ is the sum of two terms. The first is

$$-\{\Sigma_s E_s \exp[-\beta(E_s - \mu n_s)]\}/\Sigma_s \exp[-\beta(E_S - \mu n_s)] \qquad (2.48)$$

This term is $-\Sigma_s p_s E_S = -E$. The second term is

$$\Sigma_s \mu n_s \exp[-\beta(E_s - \mu n_s)]/\Sigma_s \exp[-\beta(E_s - \mu n_s)$$

which is equal to

$$\mu\Sigma_s n_s \exp[-\beta(E_s - \mu n_s)]/\Sigma_s n_s \exp[-\beta(E_S - \mu n_s) = \mu\Sigma_s p_s n_s = \mu N \qquad (2.49)$$

Finally one gets $(\partial\,\text{Ln } Z_G/\partial\beta) = -E + \mu N$ and for E one has

$$E = -\,(\partial\,\text{Ln } Z_G/\partial\beta) + \mu N = -(\partial\,\text{Ln } Z_G/\partial\beta) + (\mu/\beta)(\partial\text{Ln } Z_G/\partial\mu) \qquad (2.50)$$

when N was replaced by its value (2.46) $(1/\beta)(\partial\text{Ln } Z_G/\partial\mu)$.

Since E is equal to $-(\partial\text{Ln } Z/\partial\beta)$, we have the following relation between the grand partition function and the partition function Z of a system with the same temperature, the same mean energy and a number of particles equal to the mean number of particles of our present system.

$$(\partial \text{ Ln } Z_G/\partial\beta) = (\partial \text{Ln } Z/\partial\beta) - \mu N \tag{2.51}$$

The entropy is determined using (2.51). We saw above, in the section on the canonical ensemble that (expression 2.16)

$$S = E/T + kB \text{Ln } Z. \tag{2.16}$$

Integrating (2.51) gives

$$\text{Ln } Z_G = \text{Ln } Z - \int \mu N d\beta = \text{Ln } Z - (\mu N \beta) \tag{2.52}$$

From (2.16) one gets

$$\text{Ln } Z = [S - E/T]/kB \tag{2.53}$$

Introducing this expression of Ln Z in (2.52) gives

$$S = (E - \mu N)/T + kB \text{Ln } Z_G \tag{2.54}$$

From (2.50) $E - \mu N = -(\partial \text{Ln } Z_G/\partial\beta)$, and introducing in (2.54)

$$S = -(\partial \text{Ln } Z_G/\partial\beta)/T + kB \text{Ln } Z_G$$
$$S = -kB\beta(\partial \text{Ln } Z_G/\partial\beta) + kB \text{Ln } Z_G \tag{2.55}$$

It is not difficult to verify the relation $S = -k_B \Sigma_s p_s \text{Ln } p_s$ in analogy with the preceding cases. From the preceding results, one deduces the grand potential Ψ given by

$$\Psi(T, V, N) = -PV = E - TS - \mu N \tag{2.56a}$$
$$= -kBT \text{Ln } Z_G \tag{2.56b}$$

when its differential is $d\Psi = -SdT + PdV - Nd\mu$. To get (2.56b) one puts in (2.56a) the value of $E - \mu N = (\partial \text{Ln } Z_G/\partial\beta)$ (see 2.50) and S from (2.55). The expression (2.56) is the third link between the macroscopic thermodynamic description and the microscopic side.

2.2.3 *An example*

One considers a system in contact with a reservoir in the conditions of the grand canonical ensemble. This system is .made of particles without interaction with energies $0, e, 2e$, $3e$, etc. The grand partition function is the double sum,

$$Z_G = \Sigma_s \exp[-\beta(E_S - \mu n_s)]$$

which can be written as

$$Z_G = \Sigma_s\{\exp(\beta\mu n_s)\Sigma_s \exp(-\beta E_s)\} \tag{2.57}$$

First one performs the sum on the microstates of a system with n_s particles and then one performs the sum on the number of particles, from 0 to the infinity. The sum $\Sigma_s \exp(-\beta E_s)$ is partition function of n_s particles (with the same volume V and the same temperature T). It is not possible to use the expression (2.21), $Z = (Z_1)^N$ since now the particles are indistinguishable. In the case of the grand canonical ensemble, the particles may leave the system to enter the reservoir and vice versa as a gas. This prevents the possibility to see them as distinguishable. This fact makes the determination of the partition function difficult.

We shall consider a particular situation that we shall call the classical limit and in the next chapter we shall explain why. We suppose that particles are distributed among their possible energies or levels as follows: either one particular level is populated by one particle or it is not populated, i.e. there is no particle with this energy. When one considers a given repartition of the n_s particles in the energy levels, this defines a particular microstate. If now one changes the position of some particles but keeping the same occupied levels one has the same microstate. To take into account this fact one divides the partition function by the number of possible permutations of the particles between the occupied levels, i.e. by n_s! Thus the partition function of the n_s particles is

$$Z = (Z_1)^{n5}/(n_s!) \tag{2.58}$$

We shall find again this expression by another way in Chapter 3. The grand partition function is now

$$Z_G = \Sigma_s\{\exp(\beta\mu n_s)[\Sigma_i \exp(-\beta e_i)]^{ns}/n_s!\} \tag{2.59}$$

with $e_i = 0, e, 2e, 3e, \ldots$ etc.

The sum $\Sigma_i \exp(-\beta e_i)$ is equal to $[1 - \exp(-\beta e)]^{-1}$, this gives

$$Z_G = \Sigma_s \exp(\beta\mu n_s)[1 - \exp(-\beta e)]^{-ns}/n_s! \tag{2.60}$$

$$Z_G = \Sigma_s \{\exp(\beta\mu)[1 - \exp(-\beta e)]^{-1}\}^{ns}/n_s! \tag{2.61}$$

or

$$\text{with } x = \exp(\beta\mu)[1 - \exp(-\beta e)]^{-1}$$

$$Z_G = \Sigma_s x^{ns}/n_s! = \exp(x) \tag{2.62}$$

where we used the series development of the function $\exp(x)$. Our final result is

$$Z_G = \exp\{\exp(\beta\mu)[1 - \exp(-\beta e)]^{-1}\} \tag{2.63}$$

and

$$\text{Ln } Z_G = \exp(\beta\mu)[1 - \exp(-\beta e)]^{-1} \tag{2.64}$$

$$\Psi = -kBT \text{ Ln } Z_G = -kBT \exp(\beta\mu)[1 - \exp(-\beta e)]^{-1} \tag{2.65}$$

Now one can calculate the mean number of particles N

$$N = (1/\beta)(\partial \text{Ln } Z_G/\partial\mu) = \exp(\beta\mu)[1 - \exp(-\beta e)]^{-1} \tag{2.66}$$

Putting this expression in (2.65) gives the equation of state

$$-\Psi = PV = NkBT \tag{2.67}$$

From (2.66) one gets an expression for the chemical potential μ

$$\mu = kBT[\text{Ln}(N) + \text{Ln}(1 - \exp(-\beta e))] \tag{2.68}$$

The derivative of Ln Z_G relatively to β is following (2.50)

$$(\partial \text{Ln } Z_G/\partial\beta) = \mu N - E \tag{2.69}$$

One has

$$(\partial \text{Ln } Z_G/\partial\beta) = \mu \exp(\beta\mu)[1 - \exp(\beta e)]$$
$$- \exp(\beta\mu) e \exp(-\beta e)/[1 - \exp(-\beta e)]^{-2} \tag{2.70}$$

Taking into account of (2.56) one can write

$$(\partial \text{Ln } Z_G/\partial\beta) = \mu N - \text{Ne}\exp(-\beta e)/[1 - \exp(-\beta e)] \qquad (2.71)$$

Comparing (2.69) and (2.71) gives

$$E = \text{Ne}\exp(-\beta e)/[1 - \exp(-\beta e)] = \text{Ne}/[\exp(\beta e) - 1] \qquad (2.72)$$

The entropy can be calculated through the formula $E = TS - PV + \mu N$ giving

$$S = E/T + PV/T - \mu N/T$$

or taking account of (2.67), (2.68) and of (2.72) one gets

$$S/kBN = (e/kBT)/[\exp(e/kBT) - 1] + 1 - \text{Ln } N \\ - \text{Ln}[1 - \exp(-e/kBT)] \qquad (2.73)$$

In the next chapter, we shall show that, in the conditions we chose, the chemical potential is negative such that the entropy is always positive. It is also possible to verify that S is an increasing function of the temperature.

The condition that $\mu < 0$ gives the regime of validity of this problem. Writing (from (2.68) that

$$(\mu/kBT) = \text{Ln } N + \text{Ln}(1 - \exp(-\beta e)) < 0$$

gives

$$(\mu/kBT) = \text{Ln } N + \text{Ln}[1 - \exp(-\beta e)] = \text{Ln}[N(1 - \exp(-\beta e))] < 0$$

or $N[1 - \exp(-\beta e)] < 1$. Since N is very large, this means that $1 - \exp(-\beta e)$ is very small. In other words, $\exp(-\beta e)$ is very close to 1, i.e. βe is very small. Using the approximation $\exp(-\beta e) \approx 1 - \beta e$, the inequality $N[1 - \exp(-\beta e)] < 1$ becomes $\beta e < 1/N$ or $T > eN/k_B$. This means that the temperature is large enough and/or e (the distance between two consecutive levels) is very small. We shall see later that this situation is called the classical case.

2.3 Fluctuations

We have supposed that the mean value of the distribution energy (and other quantities) is equal to its macroscopic value although we know that this quantity fluctuates in the cases of the canonical and grand canonical ensembles. It is an important hypothesis of statistical mechanics. If fluctuations are small the energy distribution has a very sharp maximum which is the mean value of the energy and its, most probable value. One has the same thing in the case of the grand canonical ensemble about the number of the units of the system. In the next chapter on the quantum particles, we shall see that it can be easier to calculate the grand partitions function than partition function, for a given number of particles. At the end, the mean number of particles given by the grand potential is taken as the effective number of particles. This can be seen as a trick, since one wants to study a system with a well-defined number of particles by mean a method which supposes a distribution for this number. But the distribution is so sharp that the mean is effectively the number of particles.

To close this chapter, we make simple calculations of 1 "fluctuations" in order to get an a quantitative idea of the phenomenon. The goal is to estimate to what extent the values of the quantity which fluctuates are far or near the mean value. We shall begin by the energy E and one saw above, in the context of the canonical ensemble the distribution is

$$p_s = \exp[-(E_s)/kBT]/\Sigma_s \exp[-(E_s)/kBT] \tag{2.74a}$$

$$p_s = (1/Z)\exp[-(E_s)/kBT] \tag{2.74b}$$

where Z is the partition function If one writes the mean value of E as $\langle E\rangle$ and E_s the different possible values of the energy the deviations from the mean value are $[\langle E\rangle - E_5]$. But since they can be positive or negative, one takes the square of these deviations and one looks for their mean value and the fluctuations are

$$\Delta E^2 = \sum p_s[\langle E\rangle - E_s]^2 = \langle E\rangle^2 - 2\langle E_s\rangle\langle E\rangle + \langle E_s^2\rangle \tag{2.75a}$$

$$\Delta E^2 = \langle E_s^2\rangle - \langle E\rangle^2 \tag{2.75b}$$

2.3.1 Fluctuations of the energy

We begin recalling the two definition of the mean energy and that of the mean square of the energy

$$\langle E\rangle = -\sum E\exp(-\beta E)\Big/\sum \exp(-\beta E)$$

$$\langle E^2\rangle = \sum E^2\exp(-\beta E)\Big/\sum \exp(-\beta E)$$

or

$$\langle E\rangle = -(1/Z)\partial Z/\partial\beta$$
$$\langle E^2\rangle = (1/Z)\partial^2 Z/\partial\beta^2$$

Now one can write

$$\frac{\partial E}{\partial\beta} = -\frac{1}{Z}(\partial^2 Z/\partial\beta^2) + \frac{1}{z}(\partial Z/\partial\beta)^2 \qquad (2.76)$$

One has

$$\Delta E^2 = \langle E_s^2\rangle - \langle E\rangle^2 = (1/Z)\partial^2 Z/\partial\beta^2 - [(1/Z)\partial Z/\partial\beta)]^2 \qquad (2.77)$$

One sees that the derivative of $\langle E\rangle$ relatively to β is compound of two terms: (a) the derivative of $1/Z$ multiplied by $\partial Z/\partial\beta$ squared and (b) the derivative of $\partial Z/\partial\beta$ multi[plied by $1/Z$. These are exactly the two terms appearing in ΔE^2

The final result is

$$-\partial\langle E\rangle/\partial\beta = \langle E_s^2\rangle - \langle E\rangle^2$$

And since the specific heat is given by $C = \partial < E > /\partial T$ one has finally

$$\Delta E^2 = kBCT^2 \qquad (2.78)$$

We can energy to estimate the ratio

$$\sqrt{}(\Delta E^2/E) = T\sqrt{}(kBC_V)/E$$

One knows from the partition function Z^N that the energy is a linear function of N and C_V also. It results that $\sqrt{}(\Delta E^2/E)$ is proportional to $N^{-0.5}$ which is a very small number. However in certain circumstances, the fluctuations may grow and play an essential role as in second-order transitions.

2.3.2 *The fluctuations of the pressure*

In the canonical ensemble, we chose the temperature, the volume and the number of particles as the constrained quantities that define the system, Consequently, the temperature, the volume and the number of particles do not fluctuate, We shall present the results of the fluctuations of the pressure P, using the same method as above.

One has

$$P = (\partial F/\partial V)_T = (\partial E/\partial V)_S \tag{2.79}$$

$$\langle P \rangle = (1/Z) \sum (\partial E/\partial V) \exp(-\beta E) = -(1/\beta Z)(\partial Z/\partial V) \tag{2.80}$$

Finally,

$$\partial\langle P \rangle \partial V = -\beta[\langle P^2 \rangle - \langle P^2 \rangle] \tag{2.81}$$

Final remark. The first remark concerns the relation of the fluctuations with the free energy. We defined quantitatively le concept of fluctuations by the expression (2.75). In the two cases we have presented (fluctuations of the energy and the pressure) one has

$$\Delta E^2 = \langle E_s^2 \rangle - \langle E \rangle^2 = kBC_V T^2$$

$$\Delta P^2 = \langle P^2 \rangle - \langle P \rangle^2 = -kBT[\partial\langle P \rangle \partial V]$$

The quantities C_V and $\partial\langle P \rangle \partial V$ are the second derivatives of the free energy $E - TS$:

$$C_V = \frac{\partial E}{\partial T} = \left(\frac{\partial E}{\partial S}\right)\left(\frac{\partial S}{\partial T}\right) = \left(\frac{1}{T}\right)(\partial^2 F/\partial T^2)$$

$$\partial\langle P \rangle \partial V = -\partial^2 F/\partial V^2$$

This is an interesting result since this is a point of junction between thermodynamics and statistical mechanics.

$$\surd(\Delta E^2/E) = \mathrm{T}\surd(kBC_V)/E$$

One knows from the partition function Z^N that the energy is a linear function of N and C_V also. It results that $\surd(\Delta E^2/E)$ is proportional to $N^{-0.5}$ which is a very small number. However, in certain circumstances, the fluctuations may grow and play an essential role as in second-order transitions.

Chapter 3

Quantum Statistics

In this chapter, we shall calculate the general expression of the free energy from the partition function or the grand partition function of a gas, in which quantum effects are present. We shall see later that in some particular conditions the quantum effects are not important and we shall call the gas in such circumstances a classical gas or an ideal gas. As in the rest of this book, we consider the simple case of particles without interaction.

In quantum mechanics, one distinguishes between two kinds of particles: the bosons and the fermions. There are two basic differences between them. The first difference stands in the spin. The spin of a particle is $nh/2\pi$ when n takes integer values (0, 1, 2, etc.) for the bosons and half-integer values (1/2, 3/2, 5/2, etc.) for the fermions. This means that a state of a particle is also characterized by the z components s_z of the spin which can take values between $-n$ and $+n$, by steps of one $(-n, -n+1, -n+2, \ldots, n-1, i)$. Thus, the state of an isolated particle is defined by the vector linear momentum $\mathbf{p}$ and the value of s_z.

The second difference is that the fermions obey the Pauli principle: two fermions in a particular group of fermions cannot be in the same state when for the bosons such restriction does not exist. Two or more bosons can be in the same state. The fermions obey the Fermi–Dirac statistics and the bosons the Bose–Einstein statistics. Examples of fermions are the electron; the proton and the neutron and examples of bosons are the photon and the α particle.

3.1 The Partition Function and the Free Energy

When one considers a quantum gas, two different situations can happen. The number of particles may be or may be not fixed. In the first case, we indicate the number of particles by N, when in the second case this number can fluctuate. One can only speak about the mean number of particles, which is, in general, dependent of the temperature. The well-known example of a gas with a non-fixed number of particles is a gas of photons.

An isolated particle has a number of possible states that we label by a series of numbers 1, 2, 3, and so on. To each state i, corresponds an energy e_i and it is possible that two or more different states have the same energy. Since the particles are confined in a volume V, the possible energy e_i form a discontinuous series of values. We suppose that all the e_i's are positive. A microstate of the system of particles is defined by the number n_1 of particles in the state 1, n_2 particles in the state $2, \ldots, n_i$ particles in the state i, etc. The ensemble $\{n_i\}$ of the number of particles in each state is characteristic of a microstate. If one permutes two particles between the two states to which they belong, one has the same microstate since in quantum mechanics it is not possible to distinguish between two particles.

The energy of the system for a given microstate, i.e. for a given ensemble of the n_i's is

$$E_S = \Sigma_{ni} n_i e_i \tag{3.1}$$

with the following conditions: (a) For the fermions, n_i can be equal to zero or to one (impossibility to have two particles in the same state) and for the bosons n_i can take all the possible values; (b) If the number of particles is fixed, one has $N = \Sigma n_i$.

The partition function is

$$Z = \Sigma \exp[-\beta(\Sigma n_i e_i)] \tag{3.2a}$$

or

$$Z = \Sigma \exp[-\beta(n_1 e_1 + n_2 e_2 + \cdots + n_i e_i + \cdots)] \tag{3.2b}$$

when the sums are over all the microstates or over all the possible ensembles $\{n_i\}$. We write again (3.2) in a different form. Putting $z_i = \exp(-\beta e_i) < 1$, one has

$$Z = \Sigma (z_1)^{n1} (z_2)^{n2} (z_3)^{n3} \cdots (z_i)^{ni} \cdots \tag{3.3}$$

3.1.1 *Case 1. N is not fixed, fermions*

In the sum (3.3), n_i can be equal to zero or to one, and the sum (3.3) is without limit about the total number of particles. We write (3.3) summing first on n_1 (recalling that the n_i can be equal to zero or one)

$$Z = (z_1)^0 \Sigma (z_2)^{n2} (z_3)^{n3} (z_4)^{n4} \cdots$$
$$+ (z_1) \Sigma (z_2)^{n2} (z_3)^{n3} (z_4)^{n4} \cdots \tag{3.4}$$

or

$$Z = (1 + z_1) \Sigma (z_2)^{n2} (z_3^{13} (z_4)^{n4} \cdots \tag{3.5}$$

One can repeat this process for z_2 and after for z_3 and so on. We get

$$Z = (1 + z_1)(1 + z_2) \cdots (1 + z_i) \cdots \tag{3.6a}$$
$$Z = \Pi_i (1 + z_i) \tag{3.6b}$$

Now the free energy F is equal to $-k_B T \text{Ln}\ Z$ and it is

$$F = -k_B T \text{Ln}[\Pi_i (1 + z_i)] \tag{3.7a}$$

Or

$$F = -k_B T \Sigma_i \text{Ln}[1 + \exp(-\beta e_i)] \tag{3.7b}$$

3.1.2 *Case 2. N is not fixed, bosons*

We used the same procedure than for the fermions taking into account that there is no limitations on the n_i's. We begin by summing on the possible values of n_1:

$$Z = (z_1)^0 \Sigma (z_2)^{n2} (z_3)^{n3} (z_4)^{n4} + \cdots$$
$$+ (z_1) \Sigma (z_2)^{n2} (z_3)^{n3} (z_4)^{n4} + \cdots$$
$$+ (z_1)^2 \Sigma (z_2)^{n2} (z_3)^{n3} (z_4)^{n4} + \cdots$$
$$+ (z_1)^3 \Sigma (z_2)^{n2} (z_3)^{n3} (z_4)^{n4} + \cdots \tag{3.8}$$

or

$$Z = \left[1 + z_1 + (z_1)^2 + (z_1)^3 + \cdots\right] \Sigma(z_2)^{n2}(z_3)^{n3}(z_4)^{n4} \cdots \quad (3.9)$$

Since there is no limit of the possible values of the number n_1, the sum in the bracket is infinite. Since $z_i < 1$ (we recall that $e_i < 0$), the series has a limit, $(1 - z_1)^{-1}$ and one has

$$Z = (1 - z_1)^{-1}\Sigma(z_2)^{n2}(z_3)^{n3}(z_4)^{n4} \cdots \quad (3.10)$$

As above, we repeat the process for n_2, then for n_2 and so on. We obtain

$$Z = (1 - z_1)^{-1}(1 - z_2)^{-1}(1 - z_3)^{-1} \cdots \quad (3.11)$$

$$Z = \Pi_i(1 - z_i)^{-1} \quad (3.12)$$

Finally for the free energy, one has

$$F = -k_B T \Sigma_I - \mathrm{Ln}[1 - \exp(-\beta e_i)] \quad (3.13)$$

When one compares, the expressions for the fermions (3.6) and (3.7) with those for the bosons (3.12) and (3.13), one sees that the difference stands in a sign + for the fermions and – for the bosons. This permits to write in a compact form the free energy in the case of non-fixed N:

$$F = -k_B T \Sigma_i \pm \mathrm{Ln}[1 \pm \exp(-\beta e_i)] \quad (3.14)$$

and

$$Z = \Pi_i(1 \pm z_i)^{\pm 1} \quad (3.15)$$

when one puts the sign + for the fermions and the sign – for the bosons.

3.1.3 *Case 3. N is fixed, fermions*

3.1.4 *Case 4. N is fixed, bosons*

We deal with cases 3 and 4 together since, as we shall show below, the calculations are very similar to the precedent cases.

The calculations of the sum (3.3) is difficult since we have to take into account only the possible ensembles of $\{n_i\}$ for which the total number of particles is fixed, $\Sigma_i n_i = N$. We avoid this difficulty in calculating the grand partition function. But there is a price to pay, we have to introduce the chemical potential and the results are more complicated. The grand partition function Z_G is given by

$$Z_G = \Sigma_s \exp[-\beta(E_s - \mu n_s)]$$

where the sum is a double sum, on the possible sates of the system and on the number of particles. Z_G can be written as

$$Z_G = \Sigma_s \exp\{-\beta[\Sigma_i(n_i e_i) - \mu n_s]\} \tag{3.16}$$

To calculate this sum, we proceed, in principle, by steps: first, we choose a value of n_s, we distribute these n_s particles in the different states such that $\Sigma_i n_i = n_s$ and perform the sum. In a second step, we choose a different value of n_s and repeat the procedure. We can operate differently in writing Z_G in the following form:

$$Z_G = \Sigma_s \exp[-\beta(\Sigma_i n_i e_i - \mu\Sigma_i n_i)] = \Sigma_{ni} \exp[-\beta\Sigma_i n_i(e_i - \mu)] \tag{3.17}$$

or putting $z_i = \exp[-\beta(e_i - \mu)]$

$$Z_G = \Sigma_{ni}(z_1)^{n1}(z_2)^{n2}(z_3)^{n3}\cdots(z_i)^{ni}\cdots \tag{3.18}$$

which has exactly the same form than (3.3). The calculations are identical than those we did above, since in the grand canonical ensemble we perform the sum without limitation on the number of particles. We can write immediately the results from the precedent section:

$$Z_G = \Pi_i(1 \pm z_i)^{\pm 1} \tag{3.19}$$

with $z_i = \exp[-\beta(e_i - \mu)]$ and, as above, the sign $+$ is for the fermions and the sign $-$ for the bosons. From (3.19), one has

$$\text{Ln } Z_G = \Sigma_I \text{Ln}[(1 \pm z_i)^{\pm 1}] = \Sigma_i \pm \text{Ln}\{1 \pm \exp[-\beta(e_i - \mu)]\} \tag{3.20}$$

Finally, we deduce from the expressions (2.55),

$$\Psi(T, V, N) = -PV = E - TS - \mu N \tag{2.55a}$$

$$= -k_B T \text{Ln } Z_G \tag{2.55b}$$

$E - TS - \mu N = -k_B T\text{Ln } Z_G$ and the free energy $F = E - TS = -k_B T\text{Ln } Z_G + \mu N$
or

$$F = -k_B T \Sigma_i \pm \text{Ln}\,[1 \pm \exp[-\beta(e_i - \mu)] + \mu N \tag{3.21}$$

and the mean value of the number of particles N is given by

$$N = (1/\beta)(\partial \text{Ln } Z_G/\partial\mu) = \Sigma_i \exp[-\beta(e_i - \mu)] \times \{1 \pm \exp[-\beta(e_i - \mu)]\}^{-1} \tag{3.22a}$$

or multiplying the numerator and the denominator by $\exp[\beta(e_i - \mu)]$

$$N = \Sigma_i\{\exp[\beta(e_i - \mu)] \pm 1\}^{-1} \tag{3.22b}$$

The expression (3.22b) gives the possibility to calculate the chemical potential $\mu(T, V, N)$ and from the result one can calculate $F(T, V, N)$.

When one compares the expression (3.20) with (3.14) giving the free energy for the case of non-fixed N, one sees that (3.14) is a particular case of (3.21) with $\mu = 0$. Conclusion: the expression (3.20) and (3.21) are always valid with the condition $\mu = 0$ for non-fixed N.

3.2 The Energy and the Entropy

The energy is given by the expression (2.37) $E = -(\partial\text{Ln } Z_G/\partial\beta) + (\mu/\beta)(\partial\text{Ln } Z_G/\partial\mu)$ or

$$E = -(\partial\text{Ln } Z_G/\partial\beta) + \mu N.$$

One has

$$\begin{aligned}(\partial\text{Ln } Z_G/\partial\beta) &= \Sigma_i - (e_i - \mu)\exp[-\beta(e_i - \mu)][1 \pm \exp[-\beta(e_i - \mu)]^{-1} \\ &= \Sigma_i - e_i \exp[-\beta(e_i - \mu)][1 \pm \exp[-\beta(e_i - \mu)]^{-1} \\ &\quad + \mu\Sigma_i \exp[-\beta(e_i - \mu)][1 \pm \exp[-\beta(e_i - \mu)]^{-1}\end{aligned}$$

and using (3.22a)

$$-(\partial\text{Ln } Z_G/\partial\beta) = \Sigma_i e_i \exp[-\beta(e_i - \mu)]\,[1 \pm \exp[-\beta(e_i - \mu)]^{-1} - \mu N$$

and finally

$$E = \Sigma_i e_i \exp[-\beta(e_i - \mu)][1 \pm \exp[-\beta(e_i - \mu)]^{-1}$$

Multiplying the numerator and the denominator by $\exp[\beta(e_i - \mu)]$, one gets

$$E = \Sigma_i e_i \{\exp[\beta(e_i - \mu)] \pm 1\}^{-1} \quad (3.23)$$

This can be written again as $E = \Sigma_r n_r e_r$ where n_r is the average number of particles in the state r with energy e_r (we changed the labeling from n_i, number of particles in the state i at a given time to n_r, the mean number of particles in the state r for a given macrostate with energy E). One has

$$n_r = \{\exp[\beta(e_r - \mu)] \pm 1\}^{-1} \quad (3.24)$$

This expression is compatible with (3.22b) since it can be read as $N = \Sigma_r n_r$. In the case of the fermions, the n_i's can be equal to zero or to one, and as shown by (3.24) the mean numbers are always smaller than one. But in the case of the bosons, since n_r must be a positive number, one has $\exp[\beta(e_r - \mu)] > 1$ for all the possible values of e_i. This is possible only if μ is smaller than the lowest value of the e_i's. In particular, if this value is chosen to be zero, the chemical potential of a group of bosons is negative except the case of non-fixed particles when it is zero.

The expressions (3.23) and (3.24) (N is the mean number) are also valid for particles with a non-fixed number; it suffices to make $\mu = 0$.

To calculate the entropy, we use the results of the preceding chapter, expression (2.39), $S = (E - \mu N)/T + k_B \text{Ln } Z_G$. We have

$$S = (E - \mu N)/T + k_B \Sigma_i \pm \text{Ln } [1 \pm \exp[-\beta(e_i - \mu)] \quad (3.25)$$

3.3 The Classical System: The Maxwell–Boltzmann Statistics

By definition, a classical system is a system in which the quantum effects are negligible. This situation may happen if

$$\exp[\beta(e_i - \mu)] \gg 1 \quad \text{or} \quad \exp[-\beta(e_i - \mu)] \ll 1,$$

such that the expressions for the bosons and for the fermions become identical. Consequently $(\exp[\beta(e_i - \mu)] \pm 1)$ is practically equal to

$\exp[\beta(e_i - \mu)]$ and there is no difference for fermions or bosons. Since

$$n_i = \exp[-\beta(e_i - \mu)] \ll 1,$$

this is possible if practically one level is either occupied by only one particle or not occupied at all. Clearly in this case the mean number of particles in a given level is very small.

This condition must be true even for the lowest e_i that we take equal to zero. This means that $\exp(-\beta\mu) \gg 1$ or μ strongly negative (as we have supposed in the third example of the preceding chapter). Such situation occurs at high temperature associated with the property $(\partial\mu/\partial T)_{V,N} < 0$. We postpone the demonstration of this point to the next section. We precise that for a system with a non-fixed number of particles, we cannot reach such a condition (since $\mu \equiv 0$): a classical system has always a well-defined number of particles.

The condition $\exp(-\beta\mu) \gg 1$ or $\exp(\beta\mu) \ll 1$ gives the following results for the mean number of particles in the state i :

$$n_i = \exp(\beta\mu)\exp(-\beta e_i) \tag{3.26a}$$

and for the total number N (3.22b)

$$N = \exp(\beta\mu)\Sigma_i \exp(-\beta e_i) \tag{3.26b}$$

The sum $\Sigma_i \exp(-\beta e_i)$ is the partition function Z_1 of one particle and one has

$$N = \exp(\beta\mu)Z1 \tag{3.26c}$$

From (3.26a) and (3.26b) one gets the important result of the classical case

$$n_i/N = \exp(-\beta e_i)/Z_1 \tag{3.27}$$

The chemical potential is obtained from the expression (3.26c) of N

$$\beta\mu = -\text{Ln}(Z1/N) \tag{3.28}$$

$$\text{or} \quad \mu = -k_B T \text{Ln } Z1 + k_B T \text{Ln } N \tag{3.29}$$

We calculate now the Helmotz the free energy F. Since $\exp[\beta(e_i - \mu)] \gg 1$ or $\exp[-\beta(e_i - \mu)] \ll 1$ one can use, in the expression of

F(3.21), the approximation $\text{Ln}(1+x) \approx x$ for $x \ll 1$ and one gets

$$F = -k_B T \Sigma_i \exp[-\beta(e_i - \mu)] + \mu N \tag{3.30a}$$

$$F = -k_B T \exp(\beta\mu) \Sigma_i \exp(-\beta e_i) + \mu N \tag{3.30b}$$

One inserts in (3.30b) the expressions (3.27) for $\exp(\beta\mu)$ and (3.29) for μ and taking into account that

$$Z1 = \Sigma_i \exp(-\beta e_i) \text{ is the partition function of one particle,}$$
$$\beta\mu = -\text{Ln}(Z1/N)$$
$$\mu = -k_B T \text{Ln } Z1 + k_B T \text{Ln } N$$

one has

$$F = -k_B T(N/Z1)Z1 + N(-k_B T \text{Ln } Z1 + k_B T \text{Ln } N)$$

or

$$F = -k_B T(N \text{Ln } Z_1 + N - N \text{Ln } N) \tag{3.31}$$

If N is very large as we suppose, we can use the Stirling formula $\text{Ln } N! \cong N \text{Ln } N - N$ and we write F as

$$F = -k_B T[\text{Ln}(Z1)^N - \text{Ln } N!]$$

$$F = -k_B T \text{Ln}[(Z1)^N / N!] \tag{3.32}$$

We conclude that the partition function of the classical system is

$$Z = (Z1)^N / N! \tag{3.33}$$

We note that it is different from the expression (2.20) of the partition function of N-independent particles. Here, the quantity $N!$ appears in the denominator. The difference is due to the fact that in the present case, it is not possible to distinguish between the particles contrarily to the case of the expression (2.20). One has to divide by all the possible permutations of the particles.

To finish this section, we come back to the expression (3.29) giving the chemical potential. If it is strongly negative (as necessary for the ideal gas), this means that $Z_1 \gg N$. Since Z_1 is function of T and V, this condition can be translated into another condition concerning only T, V and N. We shall give explicitly this condition in the next chapter when we shall calculate the partition function of the ideal gas. This condition can be also written as $n_r \ll 1$, the mean number in one particular state is much smaller than 1.

It is usual to call such a situation of a classical case the Maxwell–Boltzmann statistics. In the present case, the particles are indistinguishable.

3.4 The Chemical Potential

The chemical potential was presented in Section 3.1 in the context of situations in which the number of particles is variable. However, in this chapter, the chemical potential is introduced in cases where the number of particles is constant. It was used as help for the application of Fermi–Dirac and Bose–Einstein statistics. It becomes an important parameter in the application of the different statistics.

We shall take one situation that will show the different behaviours of systems of bosons and system sof fermions. The goal of the calculations is to find the properties of the system as function of the temperature: energy, chemical potential and population of the lowest level).

3.4.1 *Bosons versus fermions*

One considers N particles with energy given by $e = Pn$ and $n = 1, 2, 3, \ldots$. We begin by the case of bosons. The energy is given by the expression

$$E = \Sigma_i e_i \{\exp[\beta(e_i - \mu)] - 1\}^{-1} \tag{3.23}$$

And one needs to know the chemical potential μ. It appears in the expressions of the number of particles

$$N = \Sigma_i \{\exp[\beta(e_i - \mu)] - 1\}^{-1} \tag{3.22b}$$

In order to extract the chemical potential from this expression, one has to make numeral calculations (we do not give details since only the results are of interest). We show in Fig. 3.1 the variation of the chemical potential with the temperature. In this figure, we

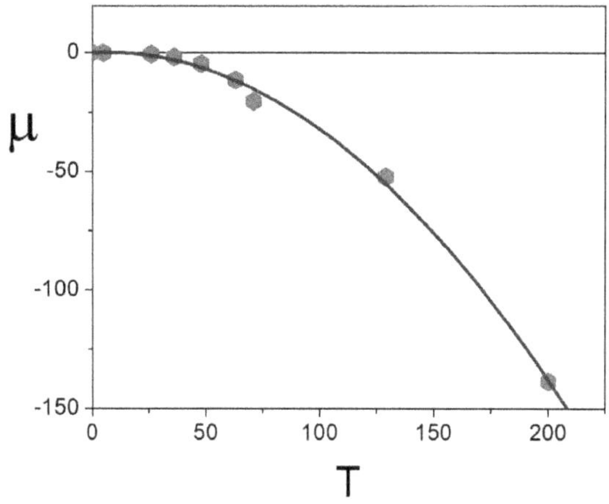

Fig. 3.1. Variations of the chemical potential of the boson gas with the temperature.

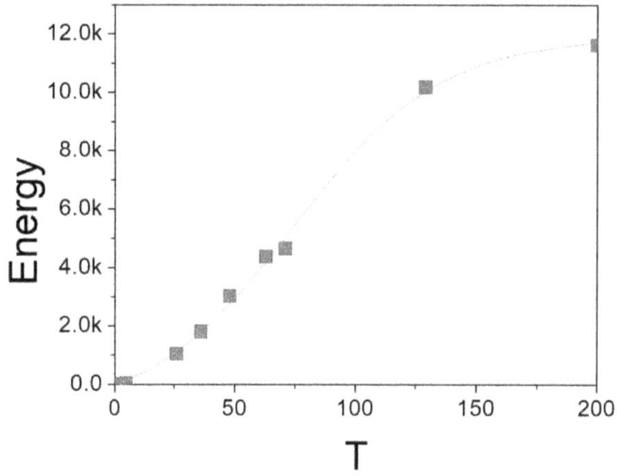

Fig. 3.2. Energy of the boson gas.

took $e = 1\,\text{eV}$ and $N = 135$, and in Fig. 3.2 the variations of the energy with T are shown. One notes an inflexion point in the energy curve which indicates that the specific heat has a maximum. One property which is particular to the bosons is the variations of the number of particles N_1 in the lower state. This is shown in Fig. 3.3 and one has a very interesting property: One can distinguish two regimes when increasing the temperature from zero where $N_1 = N$.

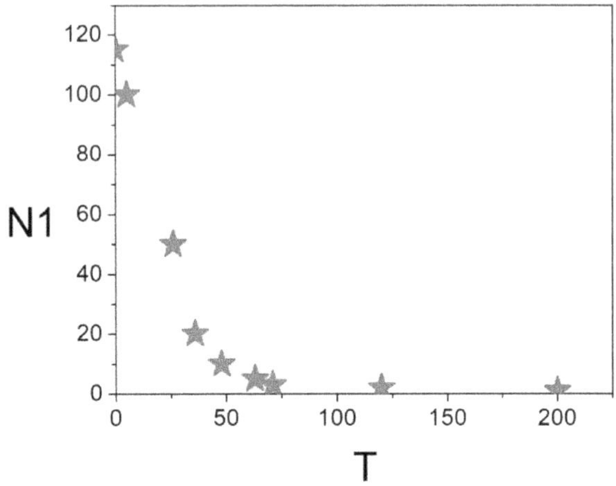

Fig. 3.3. Number of bosons in the ground state.

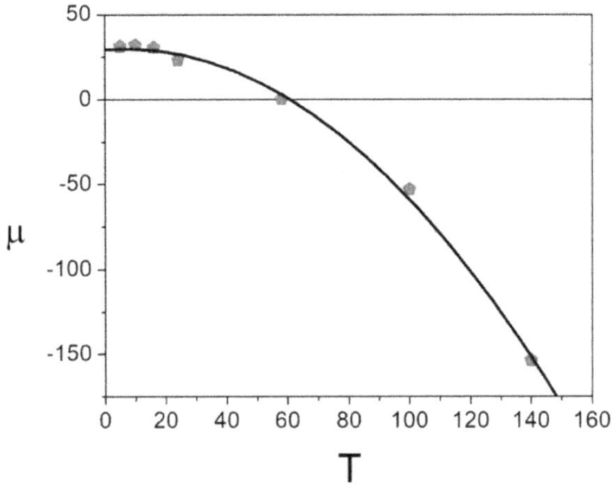

Fig. 3.4. Variation of the chemical potential of the fermions gas. Note the difference with the boson gas, Fig. 3.1.

In the first part of the curve $N_1(T)$, there is a rapid decrease of N_1 and in the second part N_1 remains small. Since we chose a small number of particles the two regimes correspond to a smooth curve. However, for a larger number of particles, the two regimes are very distinct and the lowest level becomes empty in a very small range of temperatures. In Chapter 7 on boson gas, we shall present a detailed analysis of this phenomenon called the Bose–Einstein transition.

Now one considers N fermions but one has to add some degeneracy to the energy levels in order that all the fermions could be in the lowest level. We chose $N = 500$ and degeneracy 500 so that at $T = 0$ all the fermions could be all together in the smallest ground level. We show the results of the numerical calculations: the chemical potential (Fig. 3.4), the energy (Fig. 3.5) and finally the population of the lowest level $N_1(T)$ (Fig. 3.6). It is very easy to see the striking

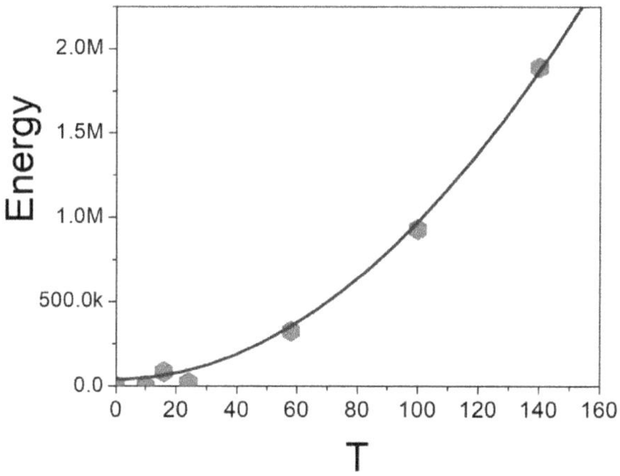

Fig. 3.5. Energy of the fermion gas.

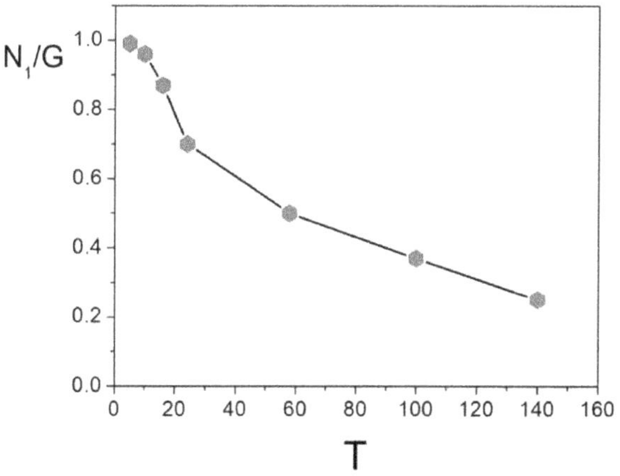

Fig. 3.6. Number of fermions in the ground state. Note also the difference with the bosons, Fig. 3.3.

differences in the behaviors of the particles due to the different statistics. In particular one can observe the variations of the particles in the ground state, The bosons come into this lowest level, such that this level, at relatively low temperature is almost full. This behavior of the fermions is different since the lowest level becomes full only progressively.

3.5 Qualitative Behavior of the Chemical Potential and Demonstration of $(\partial\mu/\partial T)_{V,N} < 0$

We begin by the expression (3.22b) giving N

$$N = \Sigma i\{\exp[\beta(ei - \mu)] \pm 1\} - 1 \tag{3.22b}$$

and derive both sides relatively to T. Since N is constant, $(dN/dT) = 0$. One writes

$$dN/dT = (dN/d\beta)(d\beta/dT) = (d\beta/dT)\Sigma i[(ei - \mu) - \beta(\partial\mu/\partial\beta)] \times \exp[\beta(ei - \mu)]/D = 0$$

when D is equal to $\{\exp[\beta(ei - \mu)] \pm 1\}2$. In the expression of $dN/dT, dT/d\beta \neq 0$ and $D \neq 0$ such that dN/dT can be equal to zero only if the sum

$$\Sigma i[(ei - \mu) - \beta(\partial\mu/\partial\beta)] \exp[\beta(ei - \mu)]$$

is equal to zero. We can decompose it into two sums that are equal:

$$\Sigma i(ei - \mu) \exp[\beta(ei - \mu)] = \beta(\partial\mu/\partial\beta)\Sigma i \exp[\beta(ei - \mu)] \tag{3.34}$$

or

$$(\partial\mu/\partial\beta) = \{\Sigma_i(ei - \mu) \exp[\beta(ei - \mu)]\}/\{\beta\Sigma i \exp[\beta(ei - \mu)]\} \tag{3.35}$$

3.5.1 *Bosons*

In the case of bosons $(ei - \mu)$ is positive since μ is smaller than all the ei's. The numerator of (3.35) is positive and the denominator also since the sum $\sum i \exp[\beta(ei - \mu)]$ is positive. One concludes that $(\partial\mu/\partial\beta) > 0$. Since $(\partial\mu/\partial T) = (\partial\mu/\partial\beta)(d\beta/dT)$ and $(d\beta/dT) = -(1/T2) < 0$, one concludes that $(\partial\mu/\partial T) < 0$.

3.5.2 *Fermions*

The case of the fermions is a little bit more complicated. We suppose, as done above, without loss of generality that all the ei's are positive and that the lowest is equal to zero. First, we note that for T going to zero, μ goes to a positive value. At $T = 0$, the system has the lowest possible energy. This means that the particles are in the lowest possible states. In the case of fermions, there is no more than one particle in each state, and consequently the states occupied by only one particle are the first N states. Recalling that the mean number in a state labeled r is $r = \{\exp[\beta(er - \mu)] + 1\} - 1$, one sees that for $\beta \to \infty \quad (T \to 0)$,

$$\text{if } er < \mu, \quad \exp[\beta(er - \mu)] \to 0 \quad \text{and} \quad n_r \to 1$$

$$\text{if } e_r > \mu, \quad \exp\left[\beta(e_r - \mu) \to \infty \quad \text{and} \quad n_r \to 0\right.$$

Since $n_r = 1$ for the N first states, we conclude that μ is larger or equal to the energy of the state with the largest occupied energy, i.e. the state $i = N$. Thus one gets the important result, $\mu(T = 0) > 0$.

Secondly, there is only one temperature $T_0 = (1/k_B\beta_0)$ for which $\mu = 0$. It is given by $N = \Sigma_i[\exp(\beta_0 e_i) + 1]^{-1}$. Thus for $T < T_0$ one has $\mu > 0$ and consequently for $T > T_0$ one has $\mu < 0$ (because T_0 is unique).[1]

Finally, for $\mu \leq 0, (\partial\mu/\partial\beta) > 0$ or $(\partial\mu/dT) < 0$ using the same reasoning that we used for the bosons. From those results, we deduce that μ is positive for $T < T_0$ and that for $T > T_0\mu$ is negative with a negative derivative $(\partial\mu/\partial T) < 0$. It is not excluded that μ may have a maximum in the region $\mu > 0$.

The conclusion is that for the bosons the chemical potential is always negative with a negative derivative relatively to the temperature when for the fermions it is the case only for T larger than a particular temperature T_0.

---

[1]We shall see in the second part that it is effectively the case for a fermions gas.

Chapter 4

The Density of States

In the definition of the partition function and the grand partition function, we insisted in specifying that the sums are over all the microstates of the system. In this chapter, we shall consider the case of a gas of particles without interaction and shall find the number $g_E(E)$ which gives, for one particle, the number of states with energy E. The partition function is $Z = \Sigma_s \exp(-\beta E_s)$ with sum over all the states. It is $\Sigma_s g_E(E) \exp(-\beta E_s)$ with sum over the energies.

Since there is no interaction, the energy of particle is the kinetic energy if the particle has a finite mass or the relativistic energy if the mass is zero. However, we shall see later (in the next chapter) that a particle (or a molecule) compound of several atoms can have also internal energy. For the time, we deal with the kinetic energy which is equal to $E = (p^2/2m)$ when p is the value of the linear momentum and m is the mass of the particle. For particles with $m = 0$ like photons the energy is relativistic, related to the momentum by $E = pc$ where c is the light velocity.

The state of a particle is defined by the vector $\mathbf{p}$ and the z component s_z of the spin. It suffices to know what is the number of the possible values of s_z and to multiply by the number of states with different vectors $\mathbf{p}$ but equal kinetic energy E to get the function $g_E(E)$, called the density of states. We begin by looking after the number: number of states with the same value of p, i.e with the same energy $E(p)$.

The second part of this chapter is devoted to the study of the ideal gas as an example of application of the density of states.

4.1 The Wave Vector

A particle in a box of dimension L_x, L_y, L_z (volume $V = L_x L_y L_z$) is described in quantum mechanics by a wave function characterized by a wave vector **k** (with component k_x, k_y, k_z)

$$\Psi(x, y, z) = \Psi_0 \sin(k_x x) \sin(k_y y) \sin(k_z z) \quad (4.1)$$

with $k_x = (n_x \pi / L_x), k_y = (n_y \pi / L_y)$ and $k_z = (n_z \pi / L_z)$ where n_x, n_y and n_z are integers $(0, 1, 2, 3$, etc). For the sake of the simplicity, one chooses $L_x = L_y = L_z = L$. This choice does not change the final results and makes the calculations simpler. One can write the squared value of the vector **k**:

$$k^2 = (\pi^2 / L^2)(n_x^2 + n_y^2 + n_z^2) \quad (4.2)$$

The relation between the linear momentum and the wave vector is $\mathbf{p} = \hbar \mathbf{k}$ and the state of the particle is defined by the vector **p** or the vector **k** or by the three numbers n_x, n_y, n_z. One defines a number n as $n^2 = n_x^2 + n_y^2 + n_z^2$ and for each ensemble of states defined by three numbers n_x, n_y, n_z having the same n, the energy E has the same value. The possible values of k are discrete but at the exception of the very low temperatures, the energy of the particles is such that the values of n are very large. For example, an electron with energy of 1 eV located in a cubic box of 1 mm side, n is of order of 10^6 and for a photon with the same energy in the same box, n is of order of 10^{19}. This justifies taking the values of n as a continuum since the two following values of n correspond to a very small difference in the energy. In such a condition, the sum giving the partition function is replaced by an integral (see below discussion on this point)

$$\mathrm{Z} = \int g_E(E) \exp(-\beta E) dE$$

This can be also be written with the help of p as

$$\mathrm{Z} = \int g_p(p) \exp[-\beta \mathrm{E}(p)] dp$$

In the case of quantum particles, one has for $\mathrm{Ln}\, Z_G$

$$\mathrm{Ln}\, Z_G = \int_{\pm} g_E(E) \{1 \pm \exp[-\beta(E - \mu)]\} dE$$

4.2 The Density of States

We have to count how many sets of the three numbers n_x, n_y, n_z located between n and $n+dn$ there are. We indicated this number by $g_n(n)dn$. We considers a three-dimensional virtual space such that a point in this space is marked by three positive numbers on the three orthogonal axis n_x, n_y, n_z. In other words, we represent a state of the particle by a point in this space. Since the numbers n_x, n_y, n_z are positive, the points corresponding to all possible states of the particles are located only in 1/8 of the total space.

The goal is to find the number of states with the same energy or to estimate the degeneracy of gas. We concluded that a state is defined by three integers and the degeneracy is given by a number n^2 equal to the sum of the squares of the three integers. In fact, one wants to know the number of boxes of size 1, in two shells with radiuses n and $n + dn$ of the 1/8 of a sphere. The calculation made in Eq. (4.3) implies that the number n is a continuous variable, supposing that the boxes are very small.

The next step is to look for the number of boxes of linear momentum Since $n = p(2L/h)$ the dimension of the ratio h/L is that of a linear momentum and this can be seen as the size of the elementary boxes.

The number of points between the two shells of radius n and $n + dn$ are the number of states $g_n(n)dn$ that we have to find. It is

$$g_n(n)dn = (1/8)4\pi n^2 dn \tag{4.3}$$

($4\pi n^2$ is the surface of one of the two spheres delimiting the shell and dn is the thickness of the shell).

We recall the relation between k and $n, k = (\pi/L)n$ and that between k and $p, p = \hbar k = (h/2\pi)k$ to get the relation between n and $p : n = p(2\text{ L}/h)$. Thus we get

$$g_n(n)dn = (1/8)4\pi n^2 dn = \pi/2[p(2L/h)]^2[(2L/h)dp] \tag{4.4a}$$

$$g_p(p)dp = 4\pi L^3 p^2 dp/h^3 \tag{4.4b}$$

Finally considering the different states with different spin component and taking that $V = L^3$ one has $g_n(n)dn = g_p(p)dp$

$$g_p(p)dp = s_z(4\pi V p^2 dp)/h^3 \tag{4.4c}$$

This is our final result, and we shall use it frequently in the following chapters. The function $g_E(E)$ is easily deduced from the equality $g_p(p)dp = g_E(E)dE$ and the relation between p and E.

Important remark. Concerning the possible energies of a particle in a box. The numbers n_x, n_y and n_z appearing in (4.2) cannot be together equal to zero. In such a case, the wave function (4.1) is null, i.e. there is no particle in the box. The smallest values of the energy are those with one of n_x, n_y, n_z is equal to 1 and the two others equal to zero (degeneracy 3 for the energy). If, in the following chapters, we shall take the energy zero as the lowest energy, it means merely that we make a change in the energy scale.

4.3 The Monatomic Ideal Gas

4.3.1 *The partition function*

The first step is to calculate the partition function $Z = (Z_1)^N/N!$ (expression (3.33) of the preceding chapter) when Z_1 is the partition function of one atom. Z_1 can be written as an integral (instead a sum) since we have shown that, for an isolated particle in a box, the different possible energies form a continuum:

$$\begin{aligned} Z_1 &= \int_0^\infty g_p(p)\exp[(-p^2/2m)/k_BT]dp \\ &= (4\pi V/h^3)\int_0^\infty p^2\exp(-\beta p^2/2m)dp \end{aligned} \tag{4.5}$$

when (4.4) was used and we supposed that $s_z = 1$. The limits of the integration are from 0 to ∞.

In order to calculate the integral (4.5), we make the change in the variable and write $x = (\beta p^2/2m)$ and one gets

$$Z_1 = [4\pi V\beta^3(2m)^{3/2}(h^{-3})]\int_0^\infty x^2\exp(-x^2)dx \tag{4.6}$$

The integral of the function $x^2\exp(-x^2)$ from 0 to ∞ is equal to $(\sqrt{\pi})/4$. Thus the final result is

$$Z_1 = V[(2\pi mk_BT)/h^2]^{3/2} \tag{4.7}$$

and the partition function is (following (3.33))

$$Z = \{V[(2\pi mk_BT)/h^2]^{3/2}\}^N/N! \tag{4.8}$$

We shall use the Stirling formula in the form $\text{Ln}(N!) = N\text{Ln}\, N - N$ and one has $N! = (N/e)^N$ where e is the basis of the natural logarithms. Inserting this expression of $N!$ (correct for large N) one has the final expression of the partition function:

$$Z = (Ve/N)^N[(2\pi mk_BT)/h^2]^{3N/2} \tag{4.9}$$

For the free energy $F = -k_BT\text{Ln}\, Z$, we get

$$F = -Nk_BT\,\text{Ln}\{(eV)/N[(2\pi mk_BT)/h^2]^{3/2}\} \tag{4.10}$$

4.3.2 *The internal energy, the entropy and the equation of state*

Now, we can calculate the three important quantities: the energy, the entropy and the equation of state.

For the energy E, we use the thermodynamic relation $E = F - T(\partial F/\partial T)$ and we get the well-known expression of the ideal gas:

$$E = (3/2)Nk_BT \tag{4.11}$$

To obtain this result, we develop the expression (4.10) of the free energy and write it as

$$F = -Nk_BT[A + (3/2)\,\text{Ln}\ T)] \tag{4.12}$$

where $A = \text{Ln}(eV/N) + 3/2\,\text{Ln}[(2\pi mk_B)/h^2]$. From $E = F - T(\partial F/\partial T)$, one gets (4.11)

The internal energy is independent of the volume and is linearly related to the temperature.

The entropy is $S = -(\partial F/\partial T)_{V,N}$ or

$$S = Nk_B\{(3/2)\text{Ln}\ T + \text{Ln}(eV/N) + 5/2 + (3/2)\,\text{Ln}[(2\pi mk_B)/h^3]\} \tag{4.13}$$

This expression is compatible with that obtained in thermodynamics since it differs only by the constant $5/2 + (3/2)\,\text{Ln}[(2\pi mk_B)/h^3$. In thermodynamics, one considers only entropy differences and this constant does play any role. But we note that this expression for S does not give $S = 0$ for $T = 0$. This is a consequence of the approximations of the classical case we made to get the partition function (no distinction between bosons and fermions).

The equation of state is $P = -(\partial F/\partial)_{T,N\Theta F}$

$$P = (Nk_BT)/V \tag{4.14}$$

This expression gives the possibility to calculate the value of the Boltzmann constant k_B. For a mole (when the number of particles is equal to the Avogadro number $N_A = 6.02 \times 10^{23}$) one has $P = (RT)/V$ (R is the ideal gas constant equal to 8.32 Joule/degree). One deduces $k_B = R/N_A = 1.38 \times 10^{-23}$ Joule/deg.

Finally, we calculate the chemical potential $\mu = (\partial F/\partial N)$. One gets

$$\mu = -k_BT\{\text{Ln}(V/N) + \text{Ln}[(2\pi mk_BT)/h^2]^{3/2}\} \tag{4.15}$$

As we mentioned above, in the classical limit, the chemical potential is negative (since as shown below $(V/N)[(2\pi mk_BT)/h^2]^{3/2} > 1$),

4.3.3 *The classical limit*

In the preceding chapter, we saw that the classical limit can be defined by the inequality $Z_1 \gg N$. Now, we are able to write explicitly this condition using the above expression of Z_1 (4.7):

$$V[(2\pi mk_BT)/h^2]^{3/2} \gg \text{N} \tag{4.16}$$

We can express this inequality in two equivalent forms. If we consider the system at constant density (N/V), we express a condition concerning the temperature:

$$kBT \gg (h^2/2\pi m)(N/V)^{2/3} \tag{4.17}$$

If we consider the system at constant temperature, the condition concerns the density

$$(N/V) \ll [(2\pi mk_BT)/h^2]^{3/2} \tag{4.18}$$

In other terms, the classical limit is reached at low density and/or at high temperature.

4.4 Some Remarks

In Chapter 2, in calculating the partition function of the classical ideal case, we did two transformations from the original series: First, we adopted the hypothesis of the equivalence series-integral and secondly, we introduced a constant Q. The role of this constant is to keep the partition function as a quantity without dimension. For the calculation of the energy this constant Q is unimportant but this introduces a constant in the expressions of the entropy, of the free energy and of the chemical potential. It is clear that, when one replaces a series by an integral, a constant must be introduced to keep the dimension of the integral identical to that of the series.

However, it is important to note that the differential density of states $g(p)dp$ does not change the dimension of the integrals.

We come back to the problem mentioned in Chapter 2 about the equivalence between the sum of a series and that of an integral. It is not a trivial problem and mathematicians have searched for different solutions. Among them, there is the formula of Euler and McLaurin. The complete formula is very complicated and we give a simplified form which is relatively precise.

One considers a series given by $\sum_0^{n-1} f(k)$ where the index k begins by 0 and goes to n. One has the following expression where $f(x)$ is a function in which x replaces the index k. The formula of Euler and McLaurin is:

$$\sum_1^{n-1} f(k) = \int_0^n f(x)dx - \frac{1}{2}[f(0) + f(n)] + \frac{1}{12}[f'(n) - f'(0)]$$

f' is the first derivative of the function f.

It is important to remark

1. This expression concerns convergent series as well as divergent series since the upper limit may be infinite or not.
2. The lower limit of k is one in the series on the right side when the lower limit of the integral is zero.

Now one compares both sides of the Euler–McLaurin expression. If one takes the integral for the complete sum of the series (from 0 to n) one makes two errors. First, the first term of the series ($k = 0$) is absent and the last term is $n - 1$. Second, this introduces two constants.

It seems that in several cases these errors do not introduce problems because frequently the number of particles in the state $k = 0$ is not very large and they do not contribute too much to the properties of the system. But we shall see in Chapter 7 that it is not the case for a gas of bosons.

Chapter 5

Some Problems

In order to have a better understanding of the concepts exposed in the preceding chapters we shall solve explicitly some problems. They are an integral part of the book and not merely exercises. It is strongly recommended to the reader not to skip this chapter.

5.1 The Quantum Harmonic Oscillator

We consider N particles with mass m, which perform a harmonic motion, in thermal contact with a reservoir at the temperature T, and we want to calculate the thermal properties of these oscillators, namely the internal energy, the specific heat and the entropy. For this purpose, we shall determine the partition function Z. We suppose that one oscillator can move in the three dimensions of the space and that the motion in each direction (x, y or z) is independent of the motion in the two others. This means that this ensemble of N oscillators in three dimensions is equivalent to $3N$ linear oscillators. We add a new assumption, namely that the positions of the oscillators are fixed and one can see them as distinguishable particles. Each oscillator is independent of the others such that one can calculate Z from the partition function of one linear oscillator Z_1 by the relation (2.20), $Z = (Z_1)^{3N}$.

The different states of the linear oscillator are characterized by their energy e

$$e = (n + 1/2)\hbar\omega \tag{5.1}$$

In (5.1), ω is the classical frequency (divided by 2π) of the oscillator. We recall that the potential energy of the oscillator is $(K\omega^2 x^2)/2$ (x is its displacement and K is a constant). The quantity N is an integer from $N = 0$ to infinity. There is one energy by state such that in the sum of the partition function there is no distinction between sum on the microstates or sum on the energies.

The one particle partition function is

$$Z_1 = \Sigma_n \exp[-\beta(n + 1/2)\hbar\omega] \tag{5.2}$$

Or

$$Z_1 = \exp(-\beta\hbar\omega/2)\Sigma_n \exp(-\beta n\hbar\omega) \tag{5.3}$$

If one writes $x = \exp[-\beta\hbar\omega]$, the sum can be written as $[1 + x + x^2 + \cdots + x^n + \cdots]$ which converges to $1/(1 - x)$ since $x < 1$. This gives for Z_1

$$Z_1 = \exp(-\beta\hbar\omega/2)/[1 - \exp(-\beta\hbar\omega)] \tag{5.4}$$

and for the total free energy of the N oscillators $F = -k_B T\text{Ln } Z$ (with $Z = (Z_1)^{3N}$)

$$F = -3Nk_B T\text{Ln } Z_1 = 3N\{\hbar\omega/2 + k_B T\text{Ln}[1 - \exp(-\beta\hbar\omega)]\} \tag{5.5}$$

To calculate the energy E, we use the relation $E = -(\partial\text{Ln } Z/\partial\beta) = -3N(\text{Ln } Z_1/\partial\beta)$. This gives

$$E = 3N\{\hbar\omega/2 + \hbar\omega[\exp(\hbar\omega/k_B T) - 1]^{-1}\} \tag{5.6}$$

The specific heat[1] C is $C = dE/dT$ and one gets after derivation

$$C = 3Nk_B(\hbar\omega/k_B T)^2 \exp(\hbar\omega/k_B T)[\exp(\hbar\omega/k_B T) - 1]^{-2} \tag{5.7}$$

It is possible to calculate the entropy following two ways. First by the expression (2.17) $S = -k_B\Sigma_s p_s\text{Ln } p_s$ and secondly by the thermodynamic relation $S = -(\partial F/\partial T)$. In the first way we shall calculate the entropy of one oscillator and since the entropy is an additive

[1]There is no distinction between the specific heart at constant volume and the specific heat at constant pressure since we supposed that the frequency of one oscillator is independent of the volume and of the pressure.

quantity, we shall multiply the result by $3N$ to get the entropy of all the oscillators. The probability p_s is given by (2.10)

$$p_s = \exp(-E_s/k_BT)/\Sigma_s \exp(-E_s/k_BT) \quad \text{or}$$

$$p_s = \exp(-E_s/k_BT)/Z_1 \tag{2.10}$$

Since the energy[2] of one oscillator is $E_s = (s+1/2)\hbar\omega$, S is given by

$$S = -k_B\Sigma_s p_s \text{Ln}\, p_s = k_B\Sigma_n\{\exp[-\beta(s+1/2)\hbar\omega]/Z_1\} \times [\beta(s+1/2)\hbar\omega + \text{Ln}\, Z_1] \tag{5.8}$$

Or

$$S = k_B\{(\beta/Z_1)\Sigma_n(s+1/2)\hbar\omega \exp[-\beta(s+1/2)\hbar\omega] + (1/Z_1)\text{Ln}\, Z_1\Sigma_n\{\exp[-\beta(s+1/2)\hbar\omega]\} \tag{5.9}$$

The first sum is minus the derivative of $\Sigma_n\{\exp[-\beta(s+1/2)\hbar\omega] = Z_1$ relatively to β and one gets for S

$$S = k_B[-\beta(\partial Z_1/\partial\beta)/Z_1 + \text{Ln}\, Z_1] \tag{5.10}$$

Since $(\partial Z_1/\partial\beta)/Z_1 = (\partial \text{Ln}\, Z_1/\partial\beta)$, we have the final expression

$$S = k_B[-\beta(\partial \text{Ln}\, Z_1/\partial\beta) + \text{Ln}\, Z_1] \tag{5.11}$$

Now from $\text{Ln}\, Z_1 = \beta\hbar\omega/2 + \text{Ln}[1-\exp(-\beta\hbar\omega)]$, it is easy to calculate S,

$$S = 3N\{(\hbar\omega/T)[\exp(\hbar\omega/k_BT)-1]^{-1} - k_B\text{Ln}[1-\exp(-\beta\hbar\omega)]\} \tag{5.12}$$

The same result is obtained from the thermodynamic relation $S = -(\partial F/\partial T)_{V,N}$.

It is interesting to look for the limits of the expressions for E, C and S in the two cases $k_BT \ll \hbar\omega$ (low-temperature limit) and $k_bT \gg \hbar\omega$ (high-temperature limit).

[2]We hope that the change in noting the quantum number n by the letter s will not disturb the reader.

5.1.1 *Low-temperature limit*

If $k_BT \ll \hbar\omega$, one has $\exp(\hbar\omega/k_BT) \gg 1$ or $\exp(-\hbar\omega/k_BT) \ll 1$. In this case, the expression of E becomes

$$E(k_BT \ll \hbar\omega) = 3N\hbar\omega[1/2 + \exp(-\hbar\omega/k_BT)] \tag{5.13}$$

which goes to $3N\hbar\omega/2$ if T goes to zero.

The specific heat is given by (5.7) and its limit for $T \to 0$ is

$$C(k_BT \ll \hbar\omega) = 3Nk_B(\hbar\omega/k_BT)^2 \exp(-\hbar\omega/k_BT) \tag{5.14}$$

which goes to zero with T. To see that, it is necessary to consider the limit of (x^2e^{-x}) for $x = \hbar\omega/k_BT$ going to infinite. This expression is the product of two quantities one going to zero (e^{-x}) and the other to infinity (x^2) but the exponential goes very fast to zero such that the product goes also to zero. It is also possible to derive (5.13) relatively to T to get (5.14).

For the limit of the entropy, we use the approximation $\text{Ln}(1-x) \approx -x$ for $|x| \ll 1$ and one gets

$$S(k_BT \ll \hbar\omega) = 3N[(\hbar\omega/T)\exp(-\hbar\omega/k_BT) + k_B \exp(-\hbar\omega/k_BT)] \tag{5.15}$$

which goes to zero with T, as expected for the entropy (recall: $(e^{-1/x})/x$ goes to zero with x).

5.1.2 *High-temperature limit*

In this case, $\hbar\omega/k_BT \ll 1$, one uses the approximation, $\exp(\pm\hbar\omega/k_BT) \approx 1 \pm \hbar\omega/k_BT$ and one has for the energy

$$E(k_bT \gg \hbar\omega) = 3Nk_BT \tag{5.16}$$

and for the specific heat

$$C(k_bT \gg \hbar\omega) = 3Nk_B \tag{5.17}$$

The entropy is the sum of two terms (see 5.12). Using the approximation $\exp(\hbar\omega/k_BT) \approx 1 + \hbar\omega/k_BT$, one sees the first term tends to $3Nk_B$. The second term $-k_B\text{Ln}[1 - \exp(-\beta\hbar\omega)]$ can be written as $k_B\text{Ln}[1 - \exp(-\beta\hbar\omega)]^{-1}$. Using again the approximation

$\exp(-\beta\hbar\omega) \approx 1 - \hbar\omega/k_BT$, this second term becomes equal to $\mathrm{Ln}(k_BT/\hbar\omega)$. Thus,

$$S(k_bT \gg \hbar\omega) = 3Nk_B[1 + \mathrm{Ln}(k_BT/\hbar\omega)] \tag{5.18}$$

One sees that E and S increase with T, when C goes to a constant.

These results were obtained first by Einstein in 1907 and have a historical importance since he was able to provide an explanation of the decrease of the specific heat of solids with T. This experimental fact was not explained by the classical statistical mechanics, which predict only the behavior given by the high-temperature limit (that we can call the classical limit). The thermal properties of solids were explained by postulating that atoms in the solids behave like harmonic oscillators with a well defined frequency. Einstein was the first to show that it is necessary to use quantum mechanics. However, his success was only qualitative since the experimental decrease of C is not given by the expression (5.14) but C is proportional to T^3. We shall see later in the second part of the book the solution to this problem.

5.2 The Polyatomic Ideal Gas

In the preceding chapter, we calculate the properties of the monatomic ideal gas, i.e. a gas compound of single atoms. In this case, the energy of one atom is its kinetic energy of translation. But in the case of molecules with several atoms, there is also, besides the kinetic energy of translation, (a) energy due to vibrations of the atoms in the molecule; (b) kinetic energy of rotation. Thus, the energy of one molecule is given by

$$E = E_t + E_{\mathrm{vib}} + E_{\mathrm{rot}} \tag{5.19}$$

where $E_t = p^2/2m$ and the one molecule partition function by

$$Z_1 = \Sigma \exp[-\beta(E_t + E_{\mathrm{vib}} + E_{\mathrm{rot}})] \tag{5.20}$$

$$Z_1 = \Sigma \exp(-\beta E_t) \exp(-\beta E_{\mathrm{vib}}) \exp(-\beta E_{\mathrm{rot}}) \tag{5.21}$$

In (5.20) and (5.21), the sum is over all the microstates, when one microstate is defined by the five numbers: the three numbers n_x, n_y, n_z of the linear momentum, the quantum number N of the vibration energy and the quantum number corresponding to the z component of the angular momentum.

We suppose that the different energies are not related, one can write Z_1 as

$$Z_1 = \Sigma_e \exp(-\beta E_t)\Sigma_{\text{vib}} \exp(-\beta E_{\text{vib}})\Sigma_{\text{rot}} \exp(-\beta E_{\text{rot}}) \qquad (5.22)$$

or

$$Z_1 = Z Z_{\text{vib}}\, Z_{\text{rot}} \qquad (5.23)$$

As an example, we shall take the case of a molecule made of two identical atoms. The energy of vibration is given by the above expression (5.1) and the energy of rotation is given by

$$E_{\text{rot}} = (\hbar^2/2I)K(K+1) \qquad (5.24)$$

I is the angular momentum of the molecule and K is a quantum number that is null or equal to a positive integer. And there are $2K+1$ states with the same energy corresponding to the z components of the angular momentum. These are the results of the quantum mechanics solution of the problem of the rotator. Z_t and Z_{vib} have been already calculated: Z_t is the one particle partition function calculated in the preceding chapter (expression (4.9)) with the difference that the mass m is now the sum of the two masses of the two atoms. Z_{vib} is given by the expression (5.4). It remains to calculate Z_{rot}.

$$Z_{\text{rot}} = \Sigma_K(2K+1)\exp[-K(K+1)\hbar^2/(2Ik_BT)] \qquad (5.25)$$

It is not possible to calculate Z_{rot} analytically and we look for approximations in the two limits $\hbar^2/2Ik_BT \ll 1$ and $\hbar^2/2Ik_BT \gg 1$. In the first case, the temperature is high enough such the important terms in the series are those with large values of K. One can replace the sum by an integral (with $K \gg 1$):

$$Z_{\text{rot}}(2Ik_BT/\hbar^2 \gg 1) = \int_0^\infty 2K \exp[-(K^2\hbar^2/2Ik_BT)]dK \qquad (5.26)$$

The integral is calculated for $K = 0$ to infinite. To calculate the integral, we make first a change in the variable, $x^2 = K^2\hbar^2/2Ik_BT$ and the integral becomes

$$Z_{\text{rot}}\ (2Ik_BT/\hbar^2 \gg 1) = (4Ik_BT/\hbar^2) \int_0^\infty x\exp(-x^2)dx \qquad (5.27)$$

The integrand is the derivative of the function $\exp(-x^2)$ multiply by $(-1/2)$ and the integral between the limits $[0, \infty]$ is equal to 1/2. The final result is

$$Z_{\text{rot}}(2Ik_BT/\hbar^2 \gg 1) = 2Ik_BT/\hbar^2 \qquad (5.28)$$

Now in the case where $2Ik_BT/\hbar^2 \ll 1$, the exponentials are small in the sum of (5.25) and one can take only the first two terms ($K = 0$ and $K = 1$):

$$Z_{\text{rot}}\ (2Ik_BT/\hbar^2 \ll 1) = 1 + 3\exp(-\hbar^2/2Ik_BT) \qquad (5.29)$$

Now, we have all the results we need to calculate the thermal properties of this diatomic ideal gas. We see that we can have different regimes following the various conditions: for the rotation $K^2\hbar^2/2Ik_BT \ll 1$ or $K^2\hbar^2/2Ik_BT \gg 1$, for the vibration $k_BT \ll \hbar\omega$ or $k_BT \gg \hbar\omega$, with the condition (4.15) $k_BT \gg (h^3/2\pi m)(N/V)^{2/3}$ of the ideal gas. We shall consider the case of the high-temperature limit in admitting that when (4.15) is satisfied, then $K^2\hbar^2/2Ik_BT \ll 1$ and $k_BT \gg \hbar\omega$. The energy and the specific heat at constant volume can be calculated using the preceding results (expression (4.11) for the energy of translation, (5.16) for the energy of vibration and expression (5.28) for the partition function of the rotation). This gives

$$E = 3/2Nk_BT + 3Nk_BT + Nk_BT = 11/2Nk_BT \qquad (5.30)$$

$$C_V = 11/2Nk_B \qquad (5.31)$$

However, in general for an ideal gas at room temperature when (4.15) is satisfied, one has $k_BT \ll \hbar\omega$ such that the contribution of the vibration to the specific heat is negligible and one has $C_V = 5/2Nk_B$ as indicated in the textbooks of Thermodynamics for a diatomic gas.

5.3 Bosons and Fermions in a Two-Level System

One considers N particles with $2M$ possible states, M states with energy equal to zero and M states with energy e. There are two possible cases: the particles are bosons or fermions. In the second case, one must have $M \geq N$ since a state cannot be occupied by more than one fermion. We want to know the thermal properties of this system of N particles. We shall see that the properties of such a system are very different depending of the nature of the particles, bosons or fermions.

To solve this problem, we do not use the partition function but the chemical potential approach. We begin by calculating the chemical potential using the expression (3.22b) that gives the number of particles

$$N = \Sigma_i\{\exp[\beta(e_i - \mu)] \pm 1\}^{-1} \tag{3.22b}$$

Again, we recall that the sum is over the states. In the present cases there are only two possible energies, but $2M$ states and N is given by

$$N = n_1 + n_2$$

$$n_1 = M[\exp(-\beta\mu) \pm 1]^{-1} \quad n_2 = M\{\exp[\beta(e - \mu)] \pm 1\}^{-1}$$

Writing

$$x = \exp(-\beta\mu) \quad \text{and} \quad A = \exp(\beta e),$$

one has an equation from which one can extract the chemical potential as a function of T and N.

$$n_1 + n_2 = M[(x \pm 1)^{-1} + (Ax \pm 1)^{-1}] = N \tag{5.32}$$

5.3.1 *The particles are bosons*

(5.32) becomes

$$M[(x - 1)^{-1} + (Ax - 1)^{-1}] = N \tag{5.33}$$

and can be written as follows:

$$M[(Ax - 1) + (x - 1)]/[(x - 1)(Ax - 1)] = N \tag{5.33b}$$

or

$$M[Ax + x - 2] = N(x - 1)(Ax - 1) \tag{5.33c}$$

or

$$(N/M)Ax^2 - (A + 1)(N/M + 1)x + (2 + N/M) = 0 \tag{5.34}$$

Before solving this equation, we shall look for solutions at $T \to \infty$ and $T \to 0$. In the first case, $A = \exp(e/k_BT) \to 1$ and (5.34) becomes

$$(N/M)x^2 - 2(N/M + 1)x + (2 + N/M) = 0 \tag{5.35}$$

This equation has two solutions that are: $x_1 = (2+N/M)(M/N)$ and $x_2 = 1$. If $x_2 = 1, n_1 = M/(x - 1)$ goes to infinity and this solution is not acceptable. Inserting the correct value of x_1 in the above expressions of n_1 and n_2 gives $n_1 = n_2 = N/2$ and $\mu = -k_BT\text{Ln}(1 + 2M/N)$.

In the second case $(T \to 0)$ $A \to \infty$, Eq. (5.34) becomes in neglecting the third term in (5.35) and dividing by A

$$(N/M)x^2 - (N/M + 1)x = 0 \tag{5.36}$$

The solution is $x = (1 + M/N)$ since $x > 0$. This gives $n_1 = N, n_2 = 0$ (since A is infinite) and $\mu = -k_BT\text{Ln}(1 + M/N)$, always negative as it stands for bosons. This analysis shows whatever the number of states and the number of particles, at low temperature all the particles are in the state 1 and the energy is null and at high temperature, the particles are equally distributed in the two levels and the energy is $E = Ne/2$.

The solution of the general problem (whatever the N and M values) is cumbersome and we chose to look for solutions in the two following cases: (1) $N = M$, (2) $N \ll M$.

1. $N = M$

Equation (5.34) becomes

$$Ax^2 - 2(A + 1)x + 3 = 0 \tag{5.37}$$

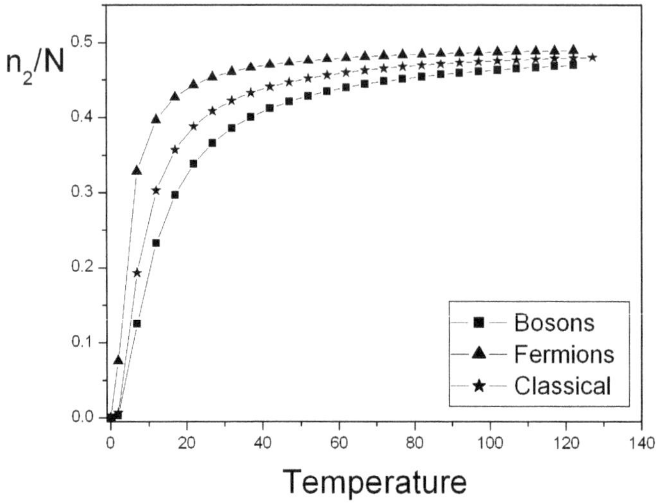

Fig. 5.1. Variation of the number of particles in the state with energy e, in the three cases: bosons, fermions and classical particles.

with two solutions $x = (1 + 1/A) \pm (1/A)\sqrt{(A^2 - A + 1)}$. We can verify that the solutions which gives the known results in the limits $T \to 0$ and $T \to \infty$, are the solutions with the sign $+$. We calculated numerically n_2, the number of particles in level 2 and we plotted in Fig. 5.1 as a function of the temperature choosing as in Chapter 1, $e/k_B = 10°K$. The energy E as a function of T has the same variation as n_2 since $E = en_2$.

2. $N \ll M$

In this case (5.34) can be written as (neglecting N/M relatively to 1 and 2)

$$(N/M)Ax^2 - (A+1)x + 2 = 0 \tag{5.38}$$

The positive solution[3] is $x = (M/2NA)\{[(A + 1) + [(A + 1)^2 - (8NA/M)]^{0.5}\}$. In the root square, one has[4] $(A + 1)^2 \gg (8NA/M)$ and the solution is with a very good approximation

[3]We chose again the positive solution for the same reason as above.

[4]We recall that A varies from 1 to infinite and consequently this inequality is always fulfilled when $M \gg N$.

$x = (M/N)(1 + 1/A)$. One remarks that in this case, x and xA are much larger than one and the expressions for n_1 and n_2 are merely

$$n_1 = (M/x) = N/(1 + 1/A) \quad n_2 = N/(1 + A)$$

The energy is

$$E = Ne[1 + \exp(\beta e)]^{-1} \tag{5.39}$$

and the chemical potential is calculated from the expression of

$$x = \exp(-\mu/k_B T) = (M/N)(1 + 1/A)$$

$$\mu = -k_B T Ln[(M/N)(1 + \exp(-\beta e)] \tag{5.40}$$

Note that μ is negative and goes to zero with T. We already note that these results were identical to the case of a two-level system in a classical situation.

5.3.2 *The particles are fermions*

The equation giving x is now (from (5.32)

$$M[(x + 1)^{-1} + (Ax + 1)^{-1}] = N \tag{5.41}$$

or (following the same steps as in the case of the bosons)

$$(N/M)Ax^2 + (A + 1)(N/M - 1)x - (2 - N/M) = 0 \tag{5.42}$$

One wants to know the limits of the populations of the two levels (for $T \to 0$ and $T \to \infty$) one transform Eq. (5.42). In the case $T \to 0$ one has $A \to \infty$ and (5.42 is written as

$$(N/M)x^2 + (N/M - 1)x = 0$$

The solution of this equation depends of the sign of $(N/M - 1)$ or if $N > M$ or if $N < M$. In the first case $(N > M)$ one gets $n_1 = M$ and $n_2 = N - M$. In the second case $(N < M)$, one gets $n_1 = N$ and $n_2 = 0$, This result is accordance with the Pauli principle.

The case $T \to \infty$ (5.42) becomes $(A \to 1)$

$$(N/M)x^2 + 2(N/M - 1)x - 2 + N/M = 0$$

$$x = (2M - N)/N$$

and the solution is $n_1 = n_2 = N/2$.

This result implies a restriction concerning the possible realization of such a system. One needs the condition $M > N/2$ because the Pauli principle.

As the case of bosons, we shall make explicit calculations for the simple cases: (1) $M = N$ and (2) $N \gg M$.

$$M = N$$

Equation (5.42) simplifies and becomes merely

$$Ax^2 - 1 = 0 \tag{5.43}$$

or

$$x = A^{-1/2} \tag{5.44}$$

and $\mu = e/2$. The chemical potential is constant and located at half distance from the two levels The final energy of this system is

$$E = Ne/[\exp(e/2k_BT) + 1]$$

and it plotted in Fig. 5.1 with the same conditions than above, i.e. $e/k_B = 10°K$. One sees that, in the case of fermions, n_2 increases much faster than for the bosons because of the Pauli principle.

2. $N \ll M$

Equation (5.43) becomes now

$$(N/M)Ax^2 - (A + 1)x - 2 = 0 \tag{5.45}$$

It is easy to show that one gets the same results for the fermions as the bosons. The conclusion is that, when $N \ll M$, there is no distinction between bosons and fermions and we consider the particles as classical particles, as in the case of the ideal gas. In Fig. 5.1, is plotted the variation of n_2 for the classical situation. One remarks that the curve of the classical particles is located between the curves of the bosons and that of fermions. The classical particles are without interaction when for the quantum particles, there are hidden interactions: repulsion for fermions and attraction for bosons.

It is worth now to give a detailed analysis of the classical situation.

5.4 Classical Particles

We want to calculate the free energy of the system in the classical case $N \ll M$. The method is general and in the classical case, it is possible to make easily explicit calculations. We know the energy

$E = Ne/[1 + \exp(\beta e)] = -(\partial \text{Ln } Z/\partial\beta)$ and the chemical potential

$$\mu = -k_B T\text{Ln}[(M/N)(1 + \exp(-\beta e)] = (\partial F/\partial N)_T$$
$$= \partial(-k_B T\text{Ln } Z)/\partial N.$$

In other words, we have the two derivatives of Ln Z from which it is possible to find the partition function Z as a function of T and N.

From the first $(\partial\text{Ln } Z/\partial\beta) = -Ne/[1 + \exp(\beta e)]$, one can write

$$\text{Ln } Z = -\int d\beta Ne/[1 + \exp(\beta e)] + K(N) \tag{5.46}$$

where $K(N)$ is an unknown function of the number of particles N. To calculate the integral, we make the change of variable $x = \beta e$ and $u = 1 + \exp(x)$. We get $du = \exp(x)dx$ or $dx = du/(u-1)$ taking into account that $\exp(x) = u - 1$. Ln Z becomes

$$\text{Ln } Z = -N\int du/[u(u-1)] + K(N) \tag{5.47}$$

Writing $1/[u(u-1)] = -1/u + 1/(u-1)$, one gets for the integral

$$\int du/[u(u-1)] = -\int du/u + \int du/(u-1) = \text{Ln}[(u-1)/u] \tag{5.48}$$

and finally for Ln Z, one has

$$\text{Ln } Z = N\text{Ln}[1 + \exp(-\beta e)] + K(N) \tag{5.49}$$

The second derivative is

$$\partial(-k_B T\text{Ln } Z)/\partial N = -k_B T\text{Ln}[(M/N)(1 + \exp(-\beta e)] \tag{5.50}$$

It can be written as

$$(\partial\text{Ln } Z/\partial N)_T = \text{Ln } M - \text{Ln } N + \text{Ln}[1 + \exp(-\beta e)] \tag{5.51}$$

This gives

$$\text{Ln } Z = -\int \text{Ln } N dN + \text{Ln } M \int dN + \text{Ln}[1+\exp(-\beta e)] \int dN + P(T) \tag{5.52}$$

where $P(T)$ is an unknown function of T.

Recalling that the integral of Ln N is ($N\text{Ln } N - N$), (5.52) can be written as

$$\text{Ln } Z = N\text{Ln}[1+\exp(-\beta e)] + N(\text{Ln } M - \text{Ln } N + 1) + P(T) \tag{5.53}$$

Comparing (5.49) and (5.53) one concludes that $P(T) \equiv 0$ and that $K(N) = N[\text{Ln}(M/N) - 1]$. The result for F is

$$F = -k_B T N\{[\text{Ln}[1+\exp(-\beta e)] + \text{Ln}(M/N) + 1\} \tag{5.54}$$

From (5.54) it is possible to calculate the entropy and its limits for $T \to 0$ and $T \to \infty$.

Finally, one can formulate the partition function Z with the help of the one-particle partition function Z_1. By definition, Z_1 is given by

$$Z_1 = M[1+\exp(-\beta e)] \tag{5.55}$$

From (5.54), the logarithm of Z is

$$\text{Ln } Z = F/(-kT) = N\{[\text{Ln}[1+\exp(-\beta e)] + \text{Ln}(M/N) + 1\} \tag{5.56}$$

or

$$\text{Ln } Z = N[\text{Ln}[1+\exp(-\beta e)] + N\text{Ln } M - N\text{Ln } N + N \tag{5.57}$$

or

$$\text{Ln } Z = N\{\text{Ln}[1+\exp(-\beta e)] + \text{Ln } M\} - N\text{Ln } N + N \tag{5.58}$$

or

$$\text{Ln } Z = \text{Ln } \{M^N[1+\exp(-\beta e)]^{N-} - N\text{Ln } N + N \tag{5.59}$$

Now we shall use the Stirling formula Ln $N! \approx N\text{Ln } N - N$ and finally

$$\text{Ln } Z = \text{Ln}\{M^N[1+\exp(-\beta e)]^N/N!\} \tag{5.60}$$

$$Z = M^N[1+\exp(-\beta e)]^N/N! = (Z_1)^N/N^! \tag{5.61}$$

One can compare the results of this section with those of the example that we took in Chapter 1 (N particles in a two-level system). In particular, the presence of $N!$ in the denominator of Z is due to the fact that, it is not possible to distinguish between the particles.

Important remark. From the preceding chapters, we see that there are four kinds of particles: Quantum particles that are indistinguishable

(1) Fermions: The grand partition function is given by $Z_G = \Pi_i(1+z_i)^{+1}$.
(2) Bosons: The grand partition function is given by $Z_G = \Pi_i(1-z_i)^{-1}$, with $z_i = \exp[-\beta(e_i-\mu)]$ N Classical particles.
(3) Distinguishable: The partition function is $Z = (Z_1)^N$.
(4) Indistinguishable: The partition function is $Z = (Z_1)^N/N!$. Z_1 is the one-particle partition function.

Particles behave classically only in particular circumstances (see the ideal gas and the preceding example).

5.5 The Magnetic Chain

We consider a chain made of N magnetic dipoles. A dipole is attached to another dipole and they form a magnetic chain. The chain is located in a plan and each dipole b can be directed along four directions $+x, -x, +y$ and $-y$. One end of the chain is fixed at the point $x = 0, y = 0$ when the other end is free and a magnetic field H is applied in the $+x$ direction. The energy of a dipole is equal to $(bH\cos\delta)$ when δ is the angle between the vector magnetic moment and the vector magnetic field. The energy of the dipole is zero if it is along the $\pm y$ directions, $e_1 = -bH$ if in the $+x$ direction and $e_2 = bH$ if it is in the $-x$ direction. Furthermore, to make the problem simple, we suppose that the state of one dipole is not influenced by the state of its two neighbors and so that two dipoles can be superposed. By this simplification, the dipoles are without interaction between them. Since the position of one dipole in the chain is well-defined, the dipoles are distinguishable units. The partition function is $Z = {Z_1}^N$. One dipole has four possible states corresponding to the four directions that it can take. The energy of these four states are $E_1 = bH, E_2 = -bH, E_3 = 0$ and $E_4 = 0$. Thus, one has for Z_1

$$Z_1 = \exp(-\beta bH) + \exp(\beta bH) + 2 = 2[1 + ch(\beta bH)] \qquad (5.62)$$

The energy is calculated with the help of the relation $E = -(\partial \text{Ln } Z/\partial\beta)$. One gets

$$E = -NbHsh(\beta bH)/[1 + ch(\beta bH)] \qquad (5.63)$$

A high temperature $(\beta \to 0), ch(\beta bH) \to 1$ and $sh(\beta bH)$ goes to zero, i.e. E goes also to zero. Now at low temperature, $(\beta \to \infty)$ and both $sh(\beta bH)$ and $ch(\beta bR)$ are very large and practically equal to $1/2 \exp(\beta bH)$. Thus, E goes to a constant value, $-NbH$.

The entropy is calculated from the expression $S = (E - F)/T = E/T + k_B \text{Ln } Z$:

$$\begin{aligned} S &= Nk_B \text{Ln}\{2[1 + ch(\beta bH)]\} \\ &\quad - (NbH/T)sh(\beta bH)/[1 + ch(\beta bH)] \end{aligned} \qquad (5.64)$$

$$\begin{aligned} S &= Nk_B\{\text{Ln } 2 + \text{Ln}[1 + ch(\beta bH)]\} \\ &\quad - Nk_B(\beta bH)sh(\beta bH)/[1 + ch(\beta bH)] \end{aligned} \qquad (5.65)$$

Now we calculate the limits of S at high and low temperature. For high $T(\beta \to 0)$, one gets $S = Nk_B Ln4$. This is a very general result for a system of N particles with a finite number of states. At low temperature, we know that S goes to zero. We shall verify this basic property of the entropy. We write S as $(x = \beta bH)$

$$S = Nk_B\{\text{Ln } 2 + \text{Ln}[1 + ch(x)]\} - Nk_B xsh(x)/[1 + ch(x)]$$

and look for the limit of S when $x \to \infty$. Putting $ch(x) \approx sh(x) \approx 1/2 \exp(x)$ for $x \to \infty$, one has

$$\begin{aligned} S &\to Nk_B\{\text{Ln } 2 + \text{Ln}[1/2 \exp(x)] - x\} \\ &= Nk_B[\text{Ln } 2 - \text{Ln } 2 + x - x] = 0 \end{aligned} \qquad (5.66)$$

The magnetization of the chain can be calculated using the magnetic free energy $M = (\partial F_M/\partial/H)$. The free energy is

$-(k_BTN)\,\mathrm{Ln}\;\; Z_1$ and one gets

$$M = Nbsh(\beta bH)/[1 + ch(\beta bH)] \tag{5.67}$$

It is also possible to calculate M through the following expression

$$M = bN(x+) - bN(x-)$$

when $N(x+)$ is the number of dipoles in the $x+$ direction and $N(x-)$ their number in the $x-$ direction. The probability $p(x+)$ for a dipole to be in the $x+$ direction is given by $\exp(\beta bH)/Z_1$ and the probability $p(x-)$ for a dipole to be in the $x-$ direction is $\exp(-\beta bH)/Z_1$. The number of dipoles in the $x+$ direction is $Np(x+)$ and the number of dipoles in the $x-$ direction is $Np(x-)$. One gets the same result (5.61).

At low temperature ($\beta \to \infty$) or/and large fields ($H \to \infty$), M tends towards Nb since $sh(\beta bH)$ and $ch(\beta bH)$ become both equal to $1/2\exp(\beta bH) \gg 1$. The magnetization has its maximum value. In this case, the chain is completely linear since all the dipoles are in the direction $x+$. However, at high temperature and low magnetic fields, M tends toward zero. This means that, in this limit, the chain has a complicated shape with the same number of dipoles in all directions. To show that recall that for $T \to \infty ch(\beta bH)$ goes to $1/2$ and $\exp(\pm\beta bH)$ goes to $1 \pm(\beta bH)$, one has

$$p(x^+) \approx (1+\beta bH)/4 \quad p(x-) \approx (1-\beta bH)/4 \quad p(y+) = p(y-) = 1/4$$

which all become equal to $1/4$ at high temperature.

As the temperature changes from zero to infinity, the chain evolves from a linear chain to a chaim with a complicated shape. The solution we present gives a simplified view of the chain. A more realistic solution is much more difficult since one has to take into account the fact that two dipoles cannot overlap.

Chapter 6

The Gas of Photons: The Black Body Radiation

When a body is heated, it is well known that it begins to emit light and the spectrum of the light emitted, i.e. the distribution of wavelengths is dependent on the temperature. If one uses the naked eye to evaluate the wavelength or the color, when the temperature of the body increases, one begins to see a red color, after appears a bright red light and further a white light. This radiation is due to the emission of electromagnetic waves or photons. The origin of the emission is the motion of the atoms of the body, which are moving because the heating. In the theory of electromagnetism, an electric charge moving in an accelerated motion emits an electromagnetic radiation. In the solid, the charged particles (electrons and nuclei) move in a disordered motion such that they emit photons. In quantum mechanics, it is taught that phenomena associated with light can be described by two complementary ways: as waves or as particles. Depending on the situation, one way is more appropriate than the other. In the present case, we shall use the particle aspect but without ignoring the other.

In this chapter, we shall consider two phenomena which apparently are not related directly: the photon gas and light emission by solids. In the case of the photon gas, i.e. photons inside a container, one wants to know, besides the regular thermodynamic properties of a gas (energy, entropy, equation of state etc.), what are the wave components of the light spectrum. On the other hand, the relevant point in the light emission of solids is precisely their light spectrum. One important case is the light emission by what it is called the black body. Such a body appears black because it absorbs all radiations whatever their wavelengths. There is a direct relation between the light spectrum of a photon gas and that of a black body

In order to be able to study precisely a photon gas, one has to define the conditions in which it takes place. For this purpose, one makes inside a body a cavity with a simple shape. Because heating of the body, the inner walls of the cavity emit photons and at the same time absorb photons. The only condition that we put on the body is that it can stand the temperature. At the equilibrium, one has a gas of photons in a well-defined volume (that of the cavity) and in a well-defined temperature. The energy of the gas is also well defined: there is no energy loss nor energy gain since the cavity is closed.

The properties of the photon gas are independent of the shape of the cavity and of the material of the walls. For the time we shall take these properties as well verified experimental facts. Below we shall give a physical argument to show that the properties of the photon gas are independent of the shape and of the material.

Now one wants to know: what is the energy of the photon gas? What is the spectrum of the radiation trapped in the cavity?

But how it is possible to answer experimentally to these questions since the gas is trapped in a close cavity without contact with the external world? To be able to make measurements on this photon gas, one makes a small aperture in the body such that a very small quantity of radiation can escape outside. This hole is so small that it does not perturb the photon gas.

The radiation spectrum of the photon gas is defined as $S(\lambda)$ giving the energy of radiation between λ and $\lambda+d\lambda$ ($d\lambda$ is a small interval of wavelength). $S(\lambda)$ goes to zero with the wavelength, reaches a maximum and decreases toward zero at large wavelength. Experimentally, one gets that E is proportional to the product VT^4. Furthermore, it was found that the product $\lambda_m T$ is constant, when λ_m is the wavelength of the maximum in the spectrum.

Before calculating the energy and the spectrum, one recalls the properties of the photon. First, as a particle it is defined by its moment $\mathbf{p}$, and its energy $E = pc$ (c is the light velocity). As a wave it has a wave vector $\mathbf{k}$, a wavelength $\lambda = 2\pi/k$ and a frequency v or $\omega = 2\pi v$. Between these quantities, one has the following relations:

$$k = 2\pi/\lambda = 2\pi v/c = \omega/c\lambda = c/v \tag{6.1}$$

The relations between the two series of properties of a photon as a particle and as a wave are

$$E = pc = h\nu = hkc/2\pi = \hbar kc \tag{6.2}$$

$$p = \hbar k \tag{6.3}$$

In (6.2) and (6.3), h is the Planck constant and $\hbar = h/2\pi$.

6.1 The Energy and the Energy Spectrum

We recall the formula for the energy that we gave in Chapter 3 (Eq. (3.23b))

$$E = \Sigma_r n_r e_r \tag{3.23b}$$

when n_r is the mean number of particles in the state r and e_r is the energy of the state r. The number n_r is given by (3.24) putting $\mu = 0$.

$$n_r = [\exp(\beta e_r) - 1]^{-1} \tag{6.4}$$

The chemical potential is null since the number of photons is not fixed. There is permanent creation and absorption of photons and their number is not fixed. The photons are bosons since its spin is 1 and thus there is no restriction of the number of photons with a given energy.

The sum (3.23b) is over the states but the sum over the energies is

$$E = \Sigma g(e_r) n_r e_r \tag{6.5}$$

where $g(e_r)$ is the density of states, the number of states with energy e_r. However, in Chapter 4 on density of states, we mentioned that the energy levels are so close that one can replace the sum (6.4) by an integral

$$E = \int_0^\infty E g(E) n(E) dE \tag{6.6}$$

with $n(E) = [\exp(\beta E) - 1]^{-1}$.

The density of states $g(p)$, as a function of the momentum p, was calculated in Chapter 4 (Eq. (4.4c)).

To transform $g(p)$ into $g(E)$, one has to recall that $E = pc$ or $p = E/c$. This gives,[1]

$$g(E)dE = (8\pi V/c^3h^3)E^2dE \tag{6.7}$$

We added a factor 2 (s_z is equal to 2 , because the two possible polarizations of the photon which is a transversal wave):

Equation (6.6) is now

$$E = (8\pi V/c^3h^3)\int_0^\infty E^3[\exp(\beta E) - 1]^{-1}dE \tag{6.8}$$

The limits of the integration are 0 and ∞. To calculate the integral, one multiplies and divides the integral by $\beta^{-4} = (k_BT)^4$. The integral becomes

$$\begin{aligned} I &= \int_0^\infty E^3[\exp(\beta E) - 1]^{-1}dE \\ &= \int_0^\infty (k_BT)^4(E/k_BT)^3[\exp(E/k_BT) - 1]^{-1}d(E/k_BT) \end{aligned}$$

or writing $x = E/k_BT$, putting $(k_BT)^4$ outside and taking the same integration limits $[0, \infty]$

$$I = (k_BT)^4\int_0^\infty x^3[\exp(x) - 1]^{-1}dE = (\pi^4/15)\,(k_BT)^4$$

$(\pi^4/15\ )$ is the value of the definite integral. Now the energy is easily calculated from Eq. (6.8): replacing the integral I by its value

$$E = (8\pi^5k^4{}_B/15h^3c^3)VT^4 \tag{6.9}$$

It is usual to write E as follows:

$$E = (4\sigma/c)VT^4 \tag{6.10}$$

[1]The spin of the photon is 1, and in principle s_z should be equal to 3. However, for a massless particle there are only two spin projections corresponding to the two possible polarizations of the light.

where $\sigma = (2\pi^5 k_B^4/15h^3c^2)$ is the Stefan constant. It is equal to $5.67\,10^{-8}$MKS and the experimental determination is in very good agreement with its theoretical formula.

The energy spectrum can be calculated from the expression (6.8) of the energy, writing it as

$$E = \int S(E)dE$$

where $S(E)dE$ is the quantity of energy between the two values E and $E + dE$. Because of the duality waveparticle, it is possible to translate the expression of $S(E)dE$ as a function of the wavelength since one has $E = pc = \hbar kc = hc/\lambda$.

From (6.8)

$$S(E)dE = \left(8\pi V/c^3h^3\right) E^3[\exp(\beta E) - 1]^{-1}dE$$

and putting $E = hc/\lambda$ and $dE = \left(hc/\lambda^2\right) d\lambda$ one gets

$$S(\lambda)d\lambda = (8\pi V hc/\lambda^5)[\exp(hc/\lambda k_B T) - 1]^{-1}d\lambda \tag{6.11a}$$

$S(\lambda)$ is the radiation spectrum of the photon gas. The quantity

$$K(\lambda) = S(\lambda)/V = (8\pi hc/\lambda^5)[\exp(hc/\lambda k_B T) - 1]^{-1} \tag{6.11b}$$

has the dimension of an energy per volume unit and per wavelength unit and can be seen as the energy density of the photon gas.

The quantity $K(\lambda)$ is plotted for several temperatures in Fig. 6.1 and these theoretical curves are in a perfect agreement with the experimental curves. The spectrum $S(\lambda)$ varies very strongly with the temperature such that we used a logarithmic scale to plot the quantity $K(\lambda)$.

To find the law of the constant product $\lambda_m T$, one derives the function $S(\lambda)$ relatively to λ and writes that the derivative is null for λ_m. First, one writes $S(\lambda) = 8\pi hc/f(\lambda)$ and one has

$$dS/d\lambda = -(8\pi hc)(df/d\lambda)/f^2$$

Writing $dS/d\lambda = 0$ is equivalent to write $df/d\lambda = 0$. The function $f(\lambda)$ is equal to

$$f(\lambda) = (\lambda^5[\exp(hc/\lambda kT) - 1]$$

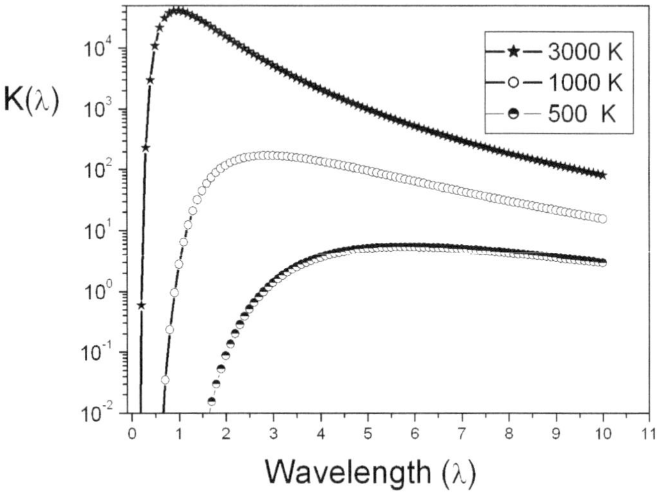

Fig. 6.1. Energy density spectrum of the photon gas as a function of the wavelength (in microns) for several temperatures. The energy spectrum is strongly dependent of the temperature.

and one has

$$df/d\lambda = 5\lambda^4[\exp(hc/\lambda_m k_B T) - 1] \\ - {\lambda_m}^3(hc/k_B T)\exp(hc/\lambda_m k_B T) = 0 \qquad (6.12a)$$

$$df/d\lambda = {\lambda_m}^4\{5[\exp(hc/\lambda_m kT) - 1] \\ - (hc/\lambda_m k_B T)\exp(hc/\lambda_m k_B T)\} = 0 \qquad (6.12b)$$

One sees that in (6.12b), one can solve the equation in the brackets taking the quantity $x = (hc/\lambda_m k_B T)$ as the unknown. One can rewrite (6.12b) as

$$df/d\lambda = {\lambda_m}^4\{[5\exp(x) - 1] - x\exp(x)\} = 0$$

Only the terms in brackets are important, one has finally the equation

$$[x\exp(x)]/[\exp(x) - 1] = 5$$

One sees that the solution must be near 5 since exp(5) is larger than 1 and consequently the left-hand becomes equal to x.

Effectively the exact solution is a transcendental number which begins by $4,96\ldots$

$$\text{Finally one can write} \quad \lambda_m T = (hc/5k_B)$$

All the preceding results were obtaining by Planck in 1909 and they were the first steps toward quantum mechanics.

6.2 The Free Energy and the Entropy

The expression (3.13) applied to the bosons gives the free energy of a gas of bosons with an undetermined number of particles.

Passing to the energy level continuum and taking into account of the density of states, one can write as

$$F = k_B T \int_0^\infty g(E)\mathrm{Ln}[1 - \exp(-\beta E)]dE \tag{6.13}$$

$g(E)$ is given by (6.7), thus

$$F = (8\pi k_B TV)/(h^3 c^3) \int_0^\infty E^2 \mathrm{Ln}[1 - \exp(-E/k_B T)]dE \tag{6.14a}$$

We repeat the trick used above to transform the integral in multiplying and dividing it by $(k_B T)^3$. This transforms the integral into a function of the group $(E/k_B T)$ alone and it becomes a definite integral.

$$\begin{aligned} F &= (8\pi k_B TV)/(h^3 c^3)(k_B T)^3 \\ &\quad \times \int_0^\infty (E/k_B T)^2 \mathrm{Ln}[1 - \exp(-E/k_B T)]d(E/k_B T) \end{aligned} \tag{6.14b}$$

This shows that F is proportional to T^4. The exact result is

$$F = -(4\sigma/3c)VT^4 \tag{6.15}$$

Formally, the negative sign comes from the fact that the logarithm in (6.14a) is negative since $[1 - \exp(-E/k_B T)]$ is smaller than 1.

From the free energy, one can calculate all the thermodynamic properties. The calculations are easy, and we give only the results that the reader can check.

The entropy

$$S = -(\partial F/\partial T)_V = (16\sigma/3c)VT^3$$

and the pressure

$$P = -(\partial F/\partial V)_T = (4\sigma/3c)T^4$$

One gets also that $PV = E/3$.

The specific heat at constant volume is $C_V = (\partial E/\partial T)_V = (16\sigma/c)VT^3$.

The mean number of photons is given by

$$N_{\text{mean}} = \Sigma_r n_r = \Sigma_r[\exp(\beta e_r) - 1]^{-1} \tag{6.16}$$

The sum is on the states and passing to the sum on the energy, one has

$$N_{\text{mean}} = \int_0^\infty g(E)[\exp(E/k_BT) - 1)]^{-1}dE \tag{6.17}$$

or using the above expression of $g(E)$

$$N_{\text{mean}} = (8\pi V)/(h^3c^3)\int_0^\infty E^2[\exp(E/k_BT) - 1)]^{-1}dE \tag{6.18}$$

Using the same trick than above, one multiplies the above expression by $(k_BT)^3$ and divides by the same quantity, thus one can write (6.18) as follows:

$$N_{\text{mean}} = (8\pi V k^3{}_B T^3)/(h^3c^3)\int_0^\infty x^2(e^x - 1)^{-1}dx \tag{6.19}$$

when the integration limits are 0 and ∞.

Taking into account that the value of the integral is approximately 7.2 the final result is

$$N_{\text{mean}} = 181(Vk^3{}_BT^3)/(h^3c^3) \tag{6.20}$$

Comparing this result with the expression of the specific heat at constant volume it is possible to see that C_v is proportional to the product $N_{\text{mean}}k_B$.

6.2.1 *The relation with the wave picture*

All the preceding results have been obtained using the particle picture of the photon gas. In a closed volume, the equivalent of a single photon is a standing wave. The problem is to count the number of standing waves with frequencies between ω and $\omega + d\omega$. Once the result is obtained, one can calculate the energy of the ensemble of the standing waves using first the equivalence between frequency and energy ($E = \hbar\omega$) and secondly the statistic weight given by $[\exp(\hbar\omega/k_B T) - 1]$.

6.3 Light Emission and Absorption of Solids: The Kirchhoff Law

We come back to the basic phenomenon of emission of light by solids when they are heated. One considers two aspects: the emission and the absorption and we shall see how they are related.

The emission power function of a heated body $B(\lambda, T)$ gives the energy of the radiation emitted by a small area of the solid surface, in a small wavelength interval $d\lambda$, per unit area and per unit time in a small solid angle $d\Omega$, as

$$dE_{em} = B(\lambda, T)(\cos\theta)d\lambda d\Omega : \tag{6.21}$$

The angle θ is the angle between the normal to the surface and the direction of the emission. $B(\lambda, T)$ is a function of the wavelength and of the temperature but also of the nature of the body. The $\cos\theta$ appears in (6.21) because one considers the emission from a small surface element dA and this surface element is seen as $(\cos\theta)dA$ in the θ direction. The emission power, expressed as a function of the wavelength and of the temperature, has the dimension of energy per unit time and unit of volume.

It is possible also to define in the same manner the radiation absorbed by a small area of the surface of the solid. It concerns only the radiation which is absorbed and not the radiation reflected or diffused by the surface. The function $M(\lambda, T)$ gives the fraction of the radiation absorbed by the body.

The Kirchhoff law states that the ratio $B(\lambda, T)/M(\lambda, T)$ depends only on λ and T and does not depend on the properties of the body.

This ratio is a universal function of λ and T. The exact expression is

$$B(\lambda, T)/M(\lambda, T) = \left(2\pi hc^2/\lambda^5\right)\left[\exp\left(hc/\lambda k_B T\right) - 1\right]^{-1} \tag{6.22a}$$

This can be written as

$$B(\lambda, T)/M(\lambda, T) = (c/2\pi)\left(8\pi V hc/\lambda^5\right)\left[\exp\left(hc/\lambda k_B T\right) - 1\right]^{-1} \tag{6.22b}$$

The Kirchhoff law can be applied to the inner walls of the cavity with a photon gas. So it not astonishing that one can write (6.22b) as

$$B(\lambda, T)/M(\lambda, T) = (c/2\pi)K(\lambda) \tag{6.23}$$

We recall that $K(\lambda)$ is the energy density per wavelength unit of the photon gas (see (6.11b)).

We shall not give a demonstration of this law but only mention an intuitive argument.

Consider the emission in a narrow range of wavelength from the wall of the cavity where the photon gas is established, it depends on the function $B(\lambda, T)$. There is also absorption by the wall of the energy from the photon gas with energy density $K(\lambda)$. Since these two processes must be equal at the equilibrium (the emitted energy flux must be equal to the absorbed energy flux) one can understand intuitively the expression (6.23) written as $B(\lambda, T) = M(\lambda, T)(c/2\pi)K(\lambda)$.

6.4 The Black Body Emission

A black body is defined as a body with perfect absorption for all wavelengths. In other words, the function $M(\lambda, T)$ is a constant equal to 1 . From (6.23), it results that the emission power of the black body is directly related to the spectrum of the photon gas.

$$B(\lambda, T) = (c/2\pi)K(\lambda)$$

This explains the equivalence between the two phenomena.

The study of light emission by solids which are not black bodies is beyond the frame of this book.

6.5 The Properties of the Photon Gas are Independent of the Shape and the Material of the Cavity

Suppose that the properties of the photon gas inside a cavity, in particular its energy, are not only dependent on the volume and the temperature but also on the shape and on the material of the body in which the cavity was made.

One considers two cavities with the same volume and the same shape at the same temperature but the two bodies are made with different materials. The two bodies with their cavities can be seen as a closed system at the temperature T. The energies of the two cavities are different because the differences in the materials. Conversely, if the energies of the two photon gases were equal, their temperature would be different.

Now one connects the two cavities by means of a tube with a very small volume giving possibility for the photons of both cavities to diffuse from one cavity to the other. After some time the energies of the two cavities are equal and consequently their temperatures are different. But one has apparition of a temperature difference in a closed system without making work. This is in contradiction with the second principle of Thermodynamics which states exactly the opposite. This demonstrates that the initial hypothesis is not correct and that the properties of a photon gas in a cavity are independent of the material.

Similarly, following the same argument, one can show that the shape of the cavity is not important.

Chapter 7

The Atomic Vibrations in Solids: Phonons

7.1 The Atomic Vibrations in Solids

In this chapter, the thermodynamics properties of solids are studied. This is a very complicated problem since atoms move in three dimensions and their motion is a collective one. An atom cannot move without influencing its neighboring atoms. In the frame of this book one can only give a simplified picture which keeps the essential of the atomic properties. In particular the specific heat of solids is well described by the model exposed in this chapter, the Debye model. We have already noted that the first model of atomic motions in solids was proposed by Einstein: each atom is seen as independent of its neighbors. Consequently, all the atoms move as harmonic oscillators with the same characteristic frequency.

We shall give briefly the properties of the motion of atoms in a crystalline solid. In a crystal the positions of atoms in the space has some regularities such that the distances between equilibrium positions of atoms are well defined.

At temperature zero all the atoms are placed at their equilibrium positions. When the temperature increases they perform small oscillations around their equilibrium positions and their amplitudes increases with the temperature. The first important point is that the atomic vibrations are the superposition of several harmonic motions given by standing waves. We take the simple example of atoms in one dimension along a line of length L. The displacements of the atoms can be perpendicular to the line (transverse motion) or along the line (longitudinal motion). A standing wave is the superposition of two propagating waves in inverse directions and opposite amplitudes. For example, the first wave is given by $A\sin(\omega T - kx)$ where A is the amplitude, ω is the frequency multiplied by 2π, T is the time, k is the wave vector $k = 2\pi/\lambda$, λ is the wavelength and x is the position along the line. The second wave is given by $-A\sin(\omega T + kx)$ and the sum is

$$A[\sin(\omega t - kx) - \sin(\omega t + kx)] = A\cos(\omega t)\sin(kx)$$

The motion of each point along the line is a harmonic motion but the amplitude depends on the position. In particular if for $x = 0$ and $x = L$ the amplitude is null if the product $kL = n\pi(n = 1, 2, 3, \ldots)$ or $\lambda = (2L/n)$: this is the case of a standing wave. In other words the length of the line is a multiple of half the wavelength.

We shall develop this simple model of a linear solid made of a line of N atoms separated the distance a, such that the length[1] of the atoms line is $L = Na$. The atoms are connected by springs with constant B. At $T = 0$ they are rest and the distance between two consecutive atoms is a. When the temperature increases the atoms perform harmonic motions under the influence of the springs. We suppose for the sake of simplicity that the atomic motions are longitudinal. One looks for the relation between the frequency ω and the wave vector k. When one atom (labeled N when N is between 1 and N) is out of its equilibrium position, two forces act on this atom, one from the atom labeled $N - 1$ and from the atom labeled $N + 1$. If we call u_n the displacement of the atom N from its equilibrium position, these forces are proportional to the differences between the displacements of two neighboring atoms, i.e. to $u_n - u_{n-1}$ and $u_n - u_{n+1}$. The equation of motion of the chain is

$$m(d^2u_n/dt^2) = -B(u_n - u_{n-1}) + B(u_{n+1} - u_n). \tag{7.1}$$

where m is the mass of one atom. We look for wave solutions given by

$$u_n = D\sin(\omega t - kna) \tag{7.2}$$

when the quantity (na) gives the position of the atom n. To check if (7.2) is solution of (7.1), one has to put it in the differential

[1]Rigorously the length of the chain is a $(N - 1)$ but since $N \gg 1$ one can take $L = Na$.

equation (7.1) and to verify for what condition the two sides of (7.1) are equal. This will give a relation between ω and k.

In place of using the solution (7.2) we shall consider the functions $D\exp[i(\omega T - kna)]$ for which the real part is (7.2). Introducing the exponential in (7.1) we shall get the condition we look for.[2] This gives

$$-m\omega^2 D\exp(i\omega t)\exp(-kna) = BD\exp(i\omega t)\{-2\exp(-kna) + \exp[-k(n+1)a] + \exp[-k(n-1)a]\} \tag{7.3a}$$

We write this equation as follows

$$-m\omega^2 D\exp(i\omega t)\exp(-kna) = BD\exp(i\omega t)\exp(-kna)[-2+\exp(-ka)+\exp(ka)] \tag{7.3b}$$

since $\exp[-k(n+1)a] = \exp(-kna)\exp(-ka)$

Using the relation $e^{ix} + e^{-ix} = 2\cos(x)$ gives $[\exp(-ka) + \exp(ka)] = 2\cos(ka)$

Introducing in (7.3b) and simplifying, one gets

$$m\omega^2 = 2B[1-\cos(ka)] \tag{7.4}$$

Since that $\cos(x) = \cos(-x)$ the expression (7.4) is also good for other solutions of (7.1) given by $u_n = D\exp[i(\omega T + kna)]$ propagating in the inverse direction as we need for standing waves solutions. Using the relation

$$\cos(x) = [1 - 2\sin^2(x/2)]$$

one has the relationship between ω and k as

$$\omega = 2(B/m)^{1/2}\sin(ka/2) \tag{7.5}$$

In Fig. 7.1, we show the graph of the frequency ω versus the wave vector for the two possible directions of the wave propagation, corresponding to positive k and to negative k. We show this relationship as a continuum but there only N possible frequencies (see below).

[2] At the end of the chapter we give the solution of (7.1) by means of trigonometric functions.

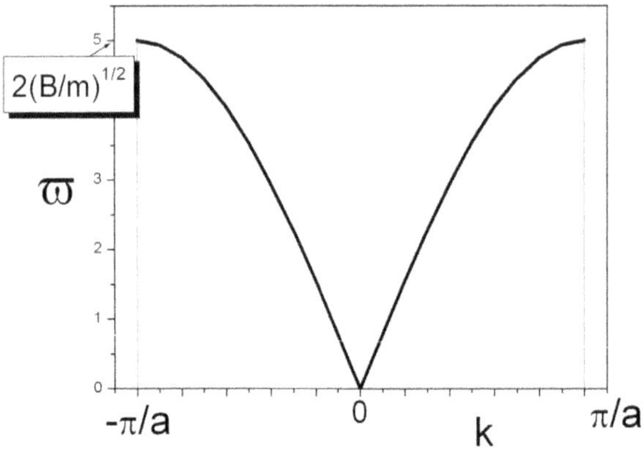

Fig. 7.1. Variation of the frequency with the wavevector for a linear atom chain.

From Fig. 7.1 and from (7.5) one can deduce important characteristics. First, there is a maximum frequency given by

$$\omega_M = 2(B/m)^{1/2}$$

Secondly corresponding to this maximum frequency, there is a maximum wave vector (in absolute value) $k = \pi/a$ and a minimum wave length $\lambda = 2\pi/k = 2a$. Thirdly, writing that the standing wave condition is that the length L is a multiple of half the wavelength, i.e. $L = d\lambda/2$ where d is an integer, one gets that the maximum d (corresponding to the smallest wave length) is equal to N. In other words, there are N possible frequencies for the chain and they are called the normal modes of the chain. Finally, we note also that only for small wave vectors, the relation between ω and k is linear using the approximation $\sin x \approx x$ for small x.

To resume: the harmonic function $D\sin(\omega T - kNa)$ is the solution of the propagating waves in the atom chain with ω related to k by (7.5) and k given by $k = \pm d(\pi/Na)(d = 1, 2, 3, \ldots, N)$.

Now we shall generalize the above results for the case of three-dimensional solids.

(A) Since the atomic motion may be transverse or longitudinal, the motion of an atom is the sum of three harmonic motions, two corresponding to the transverse waves and one to the longitudinal wave.

(B) The relation between $\acute{\omega}$ and k is not linear and since the wave velocity is given by the derivative $(d\omega/dk)$, the velocity is not independent of the frequency as in the case of electromagnetic waves. Only for the small frequencies the relationship between the frequency and the wave vector is linear resulting in independent velocity. In this range of low frequencies and large wavelengths one has $\omega = kv$ (v is the wave velocity) and since there are two types of waves one has two wave velocities. v_T (transverse wave) and v_L (longitudinal wave). These small frequencies correspond to the small wave vectors and large wavelengths which are sound waves. Effectively the sound frequencies are relatively small (say less than 20,000 Hz) and since the sound velocity is typically several thousands of meters per second in solids, it results that the wave length is much larger than the atomic distances which are order of several Angstroms. The wave vector is very small. Thus, v_T and v_L are the sound velocities in the solid.

(C) There is an upper limit for the frequencies of the waves as one saw above. This frequency is not easy to calculate and is different for the longitudinal and the transverse waves. Below we shall present the method of the Debye model to get an approximate value.

(D) There is a limited number of possible frequencies for the atomic vibrations. Since the N atoms of the solid correspond to $3N$ oscillators, the total number of possible frequencies is also $3N$ in analogy with the case of the linear atomic chain.

(E) In quantum mechanics the equivalence between waves and particles was established and we shall use it in the case of the atomic vibrations. The motion of system of the N atoms is described by the superposition of the normal modes which are standing waves. A standing wave is the wave description of a particle in a restricted volume as we saw in the case of the photon gas. Similarly to the case of electromagnetic waves where the particles are photons, one introduces a new particle, the phonon.

7.2 The Properties of the Phonons

In complete analogy with the photons, the phonons are bosons and their number is not fixed. Consequently, the chemical potential is null. The well-known relation between frequency and energy is again

$E = hv = (h\omega/2\pi)$ where h is the Planck constant. The relation between the linear momentum with the wave vector is $p = (hk/2\pi)$ as in the case of the photons.

However, there are also important differences. First there are two types of phonons, the transverse phonons corresponding to the transverse waves and the longitudinal phonons corresponding to the longitudinal waves. We recall that the photons correspond only to the transverse electromagnetic waves. Secondly their energy levels are in a finite number. Since in the wave picture there is a finite number of frequencies, one has the same properties for the energy levels. As seen above this is equivalent to say that there is a maximum possible energy E_M. Finally, the relation between the energy and the momentum is linear only for small energies. In this case, one has $E = pv_T$ for the transverse phonons and $E = pv_L$ for the longitudinal phonons. One recalls that the wave velocity is given by $d\omega/dk$ and is linear for the small values of k, i.e. the large wavelengths.

7.3 The Atomic Chain: The Low-Temperature Case

Now we present the thermal properties of the most of linear chain. We begin with the calculation of the free energy given in Chapter 3 for the case of bosons with not fixed number

$$F = -k_B T \Sigma_I - \text{Ln}[1 - \exp(-\beta e_i)] \tag{3.13}$$

or going to the energy continuum

$$F = k_B T \int_0^{E_M} g(E) \text{Ln}[1 - \exp(-\beta E)] dE \tag{7.6}$$

In (7.5) the limits of the integral are from 0 to the maximum energy E_M. One has to calculate the function $g(E)$ from the basic expression of $g_p(p)$ given in Chapter 4

$$g_p(p)dp = s_z(4\pi V p^2 dp)/h^3 \tag{4.4c}$$

However, this is not possible because the relation between E and p is not a simple function but a complicated expression, except for the low energies. This is precisely the case of the low temperatures where the energy levels occupied by the phonons are the lowest levels. One has two expressions for $g(E)$ one for the transverse phonons and one for the longitudinal ones. Using the relation $p = E/v$, one gets from (4.4c)

Longitudinal phonons: $g_L(E)dE = [4\pi V/h^3(v_L)^3]E^2dE$
Transverse phonons: $g_T(E)dE = 2[4\pi V/h^3(v_T)^3]E^2dE$

We added a factor 2 in the case of the transverse phonons as we did in the case of the photons. Finally, the total function $g(E)$ is

$$g(E) = g_T(E) + g_L(E)$$

$$g(E) = (4\pi V/h^3)[1/(v_L)^3 + 2/(v_T)^3]E^2 \tag{7.7}$$

or writing a mean sound velocity v as $3/v^3 = 1/(v_L)^3 + 2/(v_T)^3$ the final expression is

$$g(E) = (4\pi V/h^3)(3/v^3)E^2dE \tag{7.8}$$

We can write the free energy (7.6) in the following form:

$$F = (12\pi k_B TV/h^3v^3)\int_0^{E_M} E^2\text{Ln}\,[1 - \exp[-(E/k_BT)]dE \tag{7.9}$$

This expression is very similar to that of the free energy of a photon gas:

$$F = (8\pi k_B TV)/(h^3C^3)\int_0^{\infty} E^2\text{Ln}[1 - \exp[-(E/k_BT)]dE \tag{6.14a}$$

But the important difference stands in the limits of the integral: in (6.14) they are 0 and ∞ but in (7.8) they are 0 and E_M.

To calculate F (7.9) we shall use again the trick used above of multiplying and diving the expression by $(k_BT)^3$ and one gets

$$F = 12\pi(k_BT)^4V/(h^3v^3)$$
$$\times \int_0^{E_M/kT} (E/k_BT)^2\text{Ln}[1 - \exp[-E/k_BT)]d(E/k_BT) \tag{7.10}$$

with new limits for the integral: 0 and E_M/k_BT. But since we consider the low-temperature case ($E_M \gg kT$), one can take the limits as 0 and ∞. The calculation is now completely analog with that of the electromagnetic waves (see 6.14b), (7.10) becomes

$$F = 12\pi(k_BT)^4V/(h^3v^3)\int_0^{\infty} x^2\text{Ln}[1 - \exp(-x)]dx \tag{7.11}$$

where we recall that the limits of the integral are now 0 and ∞. The value of the integral is $-\pi^4/45$ and the final result is

$$F = -(4/15)\pi^5(k_BT)^4V/(h^3v^3) \tag{7.12}$$

The energy is calculated through the usual formula $E = F - T(\partial F/\partial T)$ and the specific heat at constant volume from $C_V = (\partial E/\partial T)$. The final result is

$$C_V = (16/5)[\pi^5(k_B)^4/(h^3v^3)]VT^3 \tag{7.13}$$

The variation of C_V with the temperature as T^3 is well verified for solids at low temperatures.

7.4 The Atomic Chain: The High-Temperature Case

We begin again by the free energy (7.6).

$$F = k_BT\int_0^{E_M} g(E)\mathrm{Ln}[1 - \exp(-\beta E)]dE \tag{7.6}$$

The limits of the integral are 0 and the maximum energy E_M But since we consider high temperatures ($k_BT \gg E$) we can write

$$\exp[-(E/k_BT)] \approx 1 - (E/k_BT)$$

Introducing in (7.6), we get

$$F = k_BT\int_0^{E_M} g(E)\mathrm{Ln}(E/k_BT) = k_BT \times \int_0^{E_M} g(E)[\mathrm{Ln}(E) - \mathrm{Ln}(k_BT)]dE \tag{7.14}$$

or

$$F = k_BT\int_0^{E_M} g(E)\mathrm{Ln}(E)dE - k_BT\mathrm{Ln}(k_BT) \times \int_0^{E_M} g(E)dE \tag{7.15}$$

Writing $A = \int_0^{E_M} g(E)\mathrm{Ln}(E)dE$ and $J = \int_0^{E_M} g(E)dE$, we can express F in the following form:

$$F = Ak_BT - Jk_BT\,\mathrm{Ln}(k_BT) \tag{7.16}$$

The quantities A and B depend on N and V but are independent of the temperature. We can calculate the energy $E = F - T(\partial F/\partial T)$ and the result is:

$$E = Jk_BT \tag{7.17}$$

The final step is to calculate $J = \int_0^{E_M} g(E)dE$. It is exactly the sum of all the possible states which is the number of the possible frequencies or energies. J is equal to $3N$ as we saw above. Thus, we have

$$E = 3Nk_BT \tag{7.18}$$

and

$$C_V = 3Nk_B \tag{7.19}$$

7.5 The Debye Formula

In the model of Debye it is supposed that the relation between the energy and the momentum is linear up to E_M. This means that expression for $g(E)$ is given by

$$g(E) = (4\pi V/h^3)(3/v^3)E^2dE \tag{7.8}$$

and it is supposed to be valid up to E_M. Now E_M can be calculated from the relation

$$\int_0^{E_M} g(E)dE = 3N \tag{7.20}$$

From (7.8) one has

$$\int_0^{E_M} g(E)dE = (4\pi V/h^3)(3/v^3)\int_0^{E_M} E^2dE = 3N \tag{7.21}$$

The integral is equal to $(E_M)^3/3$. This gives for E_M

$$(E_M)^3 = (3Nh^3v^3)/(4\pi V) \tag{7.22}$$

Now one can calculate the energy from the formula (3.23) of Chapter 3

$$E = \Sigma_i e_i \{\exp[\beta(e_i - \mu)] \pm 1\}^{-1} \tag{3.23}$$

Using this formula for the bosons with $\mu = 0$ and passing to the continuum limit one has

$$E = \int_0^{E_M} Eg(E)(\exp E/k_B T - 1)^{-1} dE \tag{7.23}$$

Introducing (7.8) in (7.23) gives

$$E = (4\pi V/h^3)(3/v^3) \int_0^{E_M} E^3 [\exp(E/k_B T) - 1]^{-1} dE \tag{7.24}$$

We recall that the limits of the integral are 0 and E_M. To calculate this integral we use once again the trick to transform the integral by multiplying it and dividing it by $(k_B T)^4$. This gives

$$E = (4\pi V/h^3)(3/v^3)(k_B T)^4 \times \int_0^{E_M/kT} (E/k_B T)^3 [\exp(E/k_B T) - 1]^{-1} d(E/k_B T) \tag{7.25}$$

when the limits of the integral are 0 and $(E_M/k_B T)$.

Writing $x = k_B T$ the expression for the energy is

$$E = (4\pi V/h^3)(3/v^3)(k_B T)^4 \int_0^{x_M} x^3 [\exp(x) - 1]^{-1} dx \tag{7.26}$$

The integral is a function of the temperature through the upper limit $x_M = (E_M/k_B T)$ and can be calculated numerically. It is usual to define the Debye temperature T_D by the relation

$$E_M = k_B T_D$$

and the limit x_M becomes equal to (T_D/T). It is possible to transform (7.13) in noting that $(4\pi V/h^3 v^3) = 3N/(E_M)^3$ and one gets

$$E = [9N/(E_M)^3](k_B T)^4 \int_0^{T_0/T} x^3 [\exp(x) - 1]^{-1} dx \tag{7.27}$$

or

$$E = 9Nk_BT(T/T_D)^3 \int_0^{T_D/T} x^3[\exp(x-1]^{-1}dx \tag{7.28}$$

Finally one sees that

$$E = 3Nk_BTf(T/T_D) \tag{7.29}$$

where the function

$$f(T/T_D) = 3(T/T_D)^3 \int_0^{T_D/T} x^3[\exp(x) - 1]^{-1}dx$$

is a universal function of the ratio (T/T_D), valid for all materials.

From (7.15) one can get the expression for the specific heat at a constant volume

$$C_V = (\partial E/\partial T)_V$$

$$C_V = 9Nk_B(T/T_D)^3 \int_0^{T_D/T} x^4 \exp(x)[\exp(x) - 1]^{-2}]dx \tag{7.30}$$

The passage form (7.28) to (7.30) necessitates some steps in the derivation. We put it at the end of the chapter. We recall that the limits of the integrals in (7.14, 15, 16, 17) are 0 and $x_M = T_D/T$.

C_V must be calculated numerically and we plot (C_V/Nk_B) as a function of (T/T_D) in Fig. 7.2. It is possible to show that in the low temperature limit one recovers the T^3 behavior of the specific heat. More precisely one has

$$C_V = (12\pi^4/5)Nk(T/T_D)^3 \tag{7.31}$$

and the measure of C_v at low temperatures gives a determination of T_D (of course, if the T^3 behavior is observed).

In the high temperatures limit, the Debye model gives $C_V = 3Nk_B$ as expected. Between these two limits ($T \to 0$ and $T \to \infty$), the Debye model gives a good description of C_v for materials with no too complicated structure.

Another means to check the validity of the Debye model: is through the sound velocities. We recall the definition of E_M

$$(E_M)^3 = (k_BT_D)^3 = (3Nh^3v^3)/(4\pi V) \tag{7.32}$$

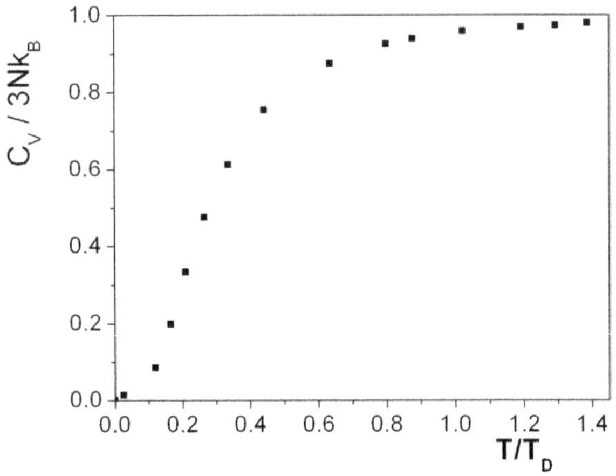

Fig. 7.2. Variation of the specific heat of the Debye model as a function of the reduced temperature (T/T_D).

Table 7.1. Debye temperature for some materials.

Materials	T_D from C_V	T_D from v
NaCl	308	320
KCl	230	246
Ag	225	216
Zn	308	305

Note: The temperatures are given in degrees Kelvin.
Source: From C. Kittel, *Introduction to Solid State Physics*. John Wiley and Sons, New York, 1956.

and one can relate the Debye temperature to the mean sound velocity v

$$T_D = v(h/k_B)(3N/4\pi V)^{1/3} \tag{7.33}$$

In Table 7.1, we give the values of the Debye temperature obtained from the specific heat and from the sound velocities for some materials. We see that the agreement is very good taking into account that the Debye model is in fact an approximation.

7.5.1 *Resolution of the differential equation* (7.1) *by means of trigonometric functions*

We recall the differential equation giving the motion of atoms in the linear atom chain

$$m(d^2u_n/dt^2) = -B(u_n - u_{n-1}) + B(u_{n+1} - u_n). \tag{7.1}$$

We look for solution as $u_n = D\sin(\omega t - kna)$. Introducing in (7.1) gives

$$\begin{aligned} -m\omega^2 \sin(\omega t - kna) &= -2B\sin(\omega t - kna) + B\sin(\omega t - k(n-1)a] \\ &\quad + B\sin[\omega t - k(n+1)a] \end{aligned} \tag{7.34}$$

One can develop the second and third term in the right-hand side of (7.34) using the formula

$$\sin(a+b) = \sin(a)\cos(b) - \cos(a)\sin(b)$$

$$\begin{aligned} \sin(\omega t - k(n-1)a] &= \sin(\omega t - kna + na) \\ &= \sin(\omega t - kna)\cos(ka) \\ &\quad + \sin(\omega t - kna)\sin(ka) \end{aligned} \tag{7.35a}$$

$$\begin{aligned} \sin(\omega t - k(n+1)a] &= \sin(\omega t - kna + na) \\ &= \sin(\omega t - kna)\cos(ka) \\ &\quad - \sin(\omega t - kna)\sin(ka) \end{aligned} \tag{7.35b}$$

The sum of these two terms is

$$2\sin(\omega t - kna)\cos(ka)$$

and introducing in (7.34) gives, after some manipulations

$$m\omega^2 = 2B[1 - \cos(ka)]$$

as we got above.

7.5.2 *Derivation of the expression* (7.30) *giving* C_V *in the Debye model*

We begin with the expression of the energy

$$E = 9Nk_BT(T/T_D)^3 \int_0^{T_D/T} x^3[\exp(x) - 1]^{-1}dx \tag{7.28}$$

and we divide the two sides by $9T_DNk_B$

$$E/(9T_DNk_B) = (T/T_D)^4 \int_0^{T_D/T} x^3[\exp(x) - 1]^{-1}dx \tag{7.36}$$

We consider the group

$$A = E/(9T_DNk_B) = y^4 \int_9^{1/y} x^3[\exp(x) - 1]^{-1}dx \tag{7.37}$$

where $y = T/T_D$. We derive A relatively to y to get the derivative relatively to T since one has

$$dA/dy = T_D(dA/dT)$$

We write

$$\int_9^{1/y} x^3(\exp x - 1)^{-1}dx = \int_9^{1/y} f(x)dx$$

and we define the function $F(x)$ such that $dF/dx = f(x)$. The integral $\int_9^{1/y} x^3(\exp x - 1)^{-1}dx$ is equal to $F(1/y) - F(0)$ and A becomes

$$A = y^4\ F(1/y)$$

since $F(0) = 0$. The derivative dA/dy is

$$dA/dy = 4y^3F(1/y) + y^4(-1/y^2)(dF/dx)_{1/y} \tag{7.38}$$

The function $F(x)$ is given by the indefinite integral

$$F(x) = \int x^3[\exp(x) - 1]^{-1}dx$$

and can be calculated by parts in writing $u = (\exp x - 1)^{-1}, dv = x^3 dx, v = x^4/4$

$$F(x) = \int u\,dv = uv - \int v\,du$$

$$F(x) = x^4/[4(\exp x - 1)] + (1/4)\int x^4 \exp(x)[\exp(x) - 1]^{-2} dx$$

Putting this expression in (7.26) and recalling that $dF/dx = x^3(\exp x - 1)^{-1}$ one gets finally

$$dA/dT = T_D(dA/dy)$$

$$dA/dT = y^3 T_D \int_9^{1/y} x^4 \exp(x)[\exp(x) - 1]^{-2} dx$$

from which one reaches the final formula

$$C_V = 9Nk_B(T/T_D)^3 \int_0^{T_D/T} x^4 \exp(x)[\exp(x) - 1]^{-2}]dx \qquad (7.30)$$

Chapter 8

The Boson Gas at Low Temperature: The Bose–Einstein Condensation

8.1 The Chemical Potential of a Boson Gas

To study the thermal properties of a gas of quantum particle one needs the knowledge of the chemical potential μ. In this chapter one considers a gas of massive bosons in a volume V with control of the temperature T.

To calculate the thermodynamic quantities, one has, following the results of Chapter 3, to determine the chemical potential as a function of the temperature through Eq. (3.22b) applied to the case of the bosons:

$$N = \Sigma_i\{\exp[\beta(e_i - \mu)] - 1\}^{-1} \tag{8.1}$$

or passing to the energy continuum

$$N = \int_0^\infty g(E)\{\exp[\beta(E - \mu)] - 1\}^{-1} dE \tag{8.2}$$

The density of states $g(E)dE$ is deduced from the expression of the density of states $g(p)dp$ obtained in Chapter 4

$$g_p(p)dp = s_z(4\pi V p^2 dp)/h^3 \tag{4.4}$$

through the relation between the momentum p and the energy E, $E = p^2/2m$ (m is the mass of an atom). One has

$$p^2 = 2mE \quad \text{and} \quad dp = mdE/p = \surd 2m^{3/2}E^{1/2}dE$$

and inserting in (4.4) one gets (taking s_z equal to one)

$$g(E)dE = (4\pi V/h^3)\sqrt{2}m^{3/2}E^{1/2}dE \tag{8.3}$$

Introducing (8.3) in (8.2) gives an expression from which on can deduce the chemical potential

$$N = (4\pi V/h^3)\sqrt{2}m^{3/2}\int_0^\infty E^{1/2}\{\exp[\beta(E-\mu)]-1\}^{-1}dE \tag{8.4}$$

$$N = [2\pi V(2m)^{3/2}/h^3]\int_0^\infty E^{1/2}\{\exp[\beta(E-\mu)]-1\}^{-1}dE \tag{8.5}$$

This is an implicit equation $N/V = F(T,\mu)$ from which one can determine μ as a function of T and the density N/V. We recall also that in the case of bosons the chemical potential must be smaller than the lowest energy; this means that μ must be negative if the lowers energy is zero. Furthermore we saw that the derivative $d\mu/dT$ is negative. Lowering the temperature makes the chemical potential smaller in absolute value and the question is: what is the limit at low temperature? Surprisingly one can see that Eq. (8.5) has the solution $\mu\cdot = 0$ for a finite temperature T_o that one can easily calculate. Effectively, putting $\mu\cdot = 0$ and $T = T_\text{o}$ in (8.5) gives

$$N = [2\pi(2m)^{3/2}V/h^3]\int_0^\infty E^{1/2}[\exp(E/k_BT_\text{o})-1]^{-1}dE \tag{8.6}$$

After multiplying the right-hand side of (8.6) by $(k_BT_\text{o})^{3/2}$, dividing $E^{1/2}$ by $(k_BT_\text{o})^{1/2}$ and dE by k_BT_o and, one gets (after rearranging the factor before the integral)

$$N = V[2\pi mk_BT_\text{o}/h^2]^{3/2}(2/\sqrt{\pi})\int_0^\infty x^{1/2}[\exp(x)-1]^{-1}dx \tag{8.7}$$

where $x = E/k_BT_\text{o}$

or

$$N/V = 2.61[2\pi mk_BT_\text{o}/h^2]^{3/2} \tag{8.9}$$

In (8.9) we replaced the integral by its value. Finally one has for T_o

$$T_\text{o} = 0.53[h^2/(2\pi mk_B)](N/V)^{2/3} \tag{8.10}$$

Thus, above T_o the chemical potential is negative but below it is likely to be positive. One concludes that something is wrong with the expression (8.2)!

The answer stands in the passage from the series (8.1) to the integral (8.2). In the series, the first term is $[\exp(-\mu/k_BT) - 1]^{-1}$ different from zero. But the function in the integral begins from zero. In fact in passing from the series to the integral, we have changed the first term with energy zero. It is just this state which becomes more and more populated when the temperature is lowered.

This interpretation finds its mathematical justification in the Euler–McLaurin formula. We recall

$$\sum_{1}^{n-1} f(k) = \int_0^n f(x)dx - \frac{1}{2}\left[f(0) + f(n) + \frac{1}{12}[f'(n) - f'(0)\right]$$

If one neglects the correction terms, one sees that in the integral the first term of the series is absent. This means merely that the integral is not equal to the sum of the series. Thus below T_o, Eq. (8.2) must be changed and written as follows, putting explicitly the first term in the sum

$$\begin{aligned} N = {} & [\exp(-\mu/k_BT) - 1]^{-1} \\ & + (4\pi V/h^3)\sqrt{2}m^{3/2} \int_0^\infty E^{1/2} \\ & \{\exp[\beta(E-\mu)] - 1\}^{-1} dE \end{aligned} \tag{8.11}$$

One can determine the behavior of μ when the temperature goes to zero. In this limit, N_1 (the number of particles in the lowest energy state) goes to N and one writes

$$N_1 = [\exp(-\mu/k_BT) - 1]^{-1} \to N \tag{8.12}$$

or

$$\exp[-\mu/k_BT] \to (N+1)/N \tag{8.13}$$

Since N is very large in a macroscopic sample, the ratio $(N+1)/N$ is very near to 1 and this mean that $\mu/k_B\ T$ is very small. In this case one develops the exponential term $\exp(-x) \approx (1-x)$ and (8.13) becomes

$$1 - \mu/k_BT \to (N+1)/N \tag{8.14}$$

and finally

$$\mu \to -k_B T/N \tag{8.15}$$

The chemical potential is thus very small at very low temperatures and it is a very good approximation to take it equal to zero below and at T_o.

To resume one writes the general expression (8.11) in the following forms:

For $T > T_o$

$$N = [2\pi(2m)^{3/2}V/h^3]\int_0^\infty E^{1/2}[\exp\{(E-\mu)/k_BT - 1]^{-1}dE \tag{8.5}$$

For $T = T_o$

$$N = [2\pi(2m)^{3/2}V/h^3]\int_0^\infty E^{1/2}[\exp(E/k_BT_o) - 1]^{-1}dE \tag{8.6}$$

For $T < T_o$

$$N = N_1 + [2\pi(2m)^{3/2}V/h^3]\int_0^\infty E^{1/2}[\exp(E/k_BT) - 1]^{-1}dE \tag{8.16}$$

The first term N_1 corresponds to the number of particles in the state with energy zero and the second term corresponds to particles number in all the other states. Remark that in (8.16) the chemical potential is taken equal to zero.

One can get from the preceding formulas (8.16) and (8.7) an expression giving N_1 as a function of T_o. Introducing the same procedure than that used above to pass from (8.6) to (8.7), one can write (8.16) as follow:

$$N = N_1 + V[2\pi mk_BT/h^2]^{3/2}(2/\sqrt{\pi})\int_0^\infty x^{1/2}[\exp(x) - 1]^{-1}dx \tag{8.17}$$

The number of particles in the state with positive energy, $N(e > 0)$ is

$$N(e > 0) = V[2\pi mk_BT/h^2]^{3/2}(2/\sqrt{\pi})\int_0^\infty x^{1/2}[\exp(x) - 1]^{-1}dx \tag{8.18}$$

One recalls the expression (8.7)

$$N = V[2\pi m k_B T_o/h^2]^{3/2}(2/\sqrt{\pi}) \int_0^\infty x^{1/2}[\exp(x) - 1]^{-1} dx \quad (8.7)$$

From (8.7) one has

$$(2/\sqrt{\pi}) \int_0^\infty x^{1/2}[\exp(x) - 1]^{-1} dx = N/[V(2\pi m k_B T_o/h^2)^{3/2}]$$

inserting in (8.17) one has $N = N_1 + N(T/T_o)^{3/2}$ and finally

$$N_1/N = 1 - (T/T_o)^{3/2} \quad (8.19)$$

The temperature T_o can be seen as the temperature for which the "condensation" of the particles into the lowest state begins. For $T \to 0$, the number of particles in this state goes to N and for $T \geq T_o$, this number is null. This last point, namely that for $T > T_o$ the lowest level is partially unoccupied is specific to the bosons as shown in Chapter 3.

This phenomenon was predicted by Einstein using a method developed by Bose, it receives the name of the Bose–Einstein condensation.

8.2 The Energy, the Specific Heat, the Free Energy and the Entropy

We consider the case $T \leq T_o$ which is simpler because the chemical potential is null. The energy is given by

$$E = \int_0^\infty E g(E) dE = \int_0^\infty E(4\pi V/h^3)\sqrt{2}m^{3/2}E^{1/2} \times [\exp(E/k_B T) - 1]^{-1} dE \quad (8.20)$$

One uses again the trick of transforming the integrals by dividing and multiplying this expression by $(k_B T)^{5./2}$ and rearranging the numerical factor one has

$$E = V[2\pi m/h^2]^{3/2}(2/\sqrt{\pi})(k_B T)^{5./2} \int_0^\infty x^{3/2}[\exp(x) - 1]^{-1} dx \quad (8.21)$$

One recalls the expression giving N (8.7)

$$N = V[2\pi m k_B T_o/h^2]^{3/2}(2/\sqrt{\pi})\int_0^\infty x^{1/2}[\exp(x)-1]^{-1}dx \quad (8.7)$$

Dividing (8.21) by (8.7) term by term gives

$$E/N = [T^{5/2}/(T_o)^{3.2}]\int_0^\infty x^{3/2}[\exp(x)-1]^{-1}dx \Big/ \left\{\int_0^\infty x^{1/2}[\exp(x)-1]^{-1}dx\right\} \quad (8.22)$$

The ratio of the two integrals is about 0.77 thus one gets finally

$$E = 0.77 N k_B [T^{5/2}/(T_o)^{3.2}] \quad (8.23)$$

It is possible to express E as a function of V and T introducing in (8.23) the values of T_o given by (8.10)

$$E = (2\pi m/h^2)^{3/2} V (k_B T)^{5/2} = AVT^{5/2} \quad (8.24)$$

where A is equal to $(2\pi m/h^2)^{3/2}(k_B)^{5/2}$. Now one can calculate the specific heat at constant volume C_V, the entropy and the free energy F:

$$C_V = (\partial E/\partial T)V = 5/2\, AVT^{3/2} \quad (8.25)$$

$$S = \int (C_V/T)dT = 5/3\, AVT^{3/2} \quad (8.26)$$

$$F = E - TS = -2/3\, AVT^{5/2} = -2/3\, E \quad (8.27)$$

From (8.27) one deduces the pressure

$$P = -(\partial F/\partial V)_T = 2/3\, AT^{5/2} \quad (8.28)$$

One remarks the very particular behavior of the boson gas in the condensation regime. The energy, the free energy and the entropy do not depend on the number of particles It is merely the consequence of the fact that the chemical potential is zero. The pressure is independent of the volume.

Immediately above T_o the calculations become cumbersome and we do not present them. And at temperatures well above T_o the gas behaves as a classical ideal gas with energy $3/2 N k_B T$.

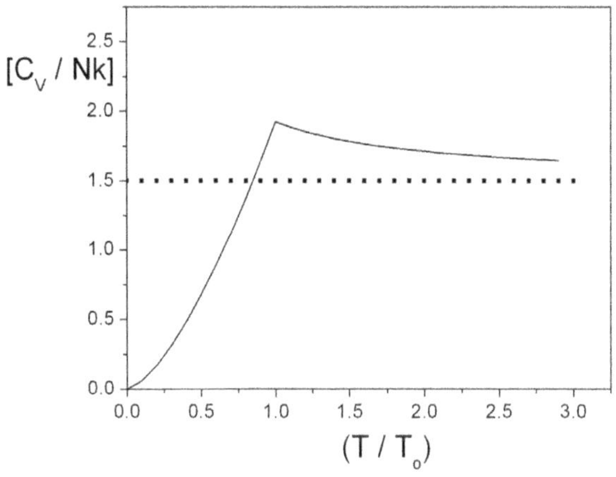

Fig. 8.1. Variation of the specific heat with the reduced temperature showing the beak in the slope at $T = T_o$.

We shall give a qualitative picture of the specific heat C_V, For $T \leq T_o$ one comes back to the expression (8.23) of the energy to calculate again the specific hast at constant volume

$$C_V = (\partial E/\partial T)_V = 1.925 N k_B (T/T_o)^{3/2}$$

Well above T_o one has the classical value $C_V = 3./2Nk_B$. However, for $T = T_o C_V$ is equal to $1.925Nk_B$, larger than the classical value. It is possible to show that the specific heat is continuous at T_o and thus for $T > T_o C_V$ begins to decrease from the value $1.925Nk_B$ to the value $3/2Nk_B$. In Fig. 8.1, one shows schematically the variation of C_V with T.

8.3 Experimental Confirmations

The Bose–Einstein condensation was predicted in 1925 and at this time there was no experimental evidence of such a phenomenon. The first proposition for a possible experimental realization was made by London in 1938 about the properties of liquid helium. This element is a gas at room temperature and it remains liquid even at very low temperature. Particular properties appear at the temperature of 2.3°K when it losses its viscosity. This state is called a superfluid state. The idea of London was to consider this transition as a

Bose–Einstein condensation. If one applies the formula (8.10) to calculate the Bose–Einstein condensation, one obtains T_o about 3.4°K. It is not exactly the good temperature, but it is not too far. In order to understand the proposal of London, one considers the two fluid model proposed later by Tisza. In this model, the helium liquid in its superfluid state is compound of two fluids. In the first the atoms form a normal liquid and in the other the atoms form the superfluid liquid. The atoms in this superfluid state are those which condensate in the lowest energy level as in the Bose–Einstein condensation. However, the application of the Bose–Einstein condensation to helium presents several difficulties since the theory considers atoms in their gaseous state when they do not have any interaction between them contrarily to liquid helium.

The problem in an experimental realization of the Bose–Einstein condensation is as follows. From the expression (8.10) of T_o it possible to see that the condensation can occur only at low temperatures when all the elements are solid (at the exception of helium which remains liquid) If the number of atoms is reduced to a small number (for example, 10^6 atoms or less) it is possible to kept them in a gaseous state. But the condensation temperature becomes very low (below 10^{-60} K). The technical difficulties are to get a gas of small number of atoms at a very low temperature. Only very recently solution to this problem was found and the Bose–Einstein condensation was observed. These are very sophisticated techniques first to cool the atoms (like Rubidium atoms) by laser and evaporation and secondly to keep the group of atoms far from the wall of the container (to avoid heating of the atoms) by magnetic interaction. The atoms in the condensate state are in their lowest kinetic energy stare and consequently are almost stationary.

Chapter 9

The Gas of Fermions: Electrons in Metals and in Semiconductors; Neutron Stars

In this chapter, we consider a gas of fermions in three particular situations. In the first, we present a simple model of metals described by a gas of N electrons in a box of volume V. The model is used to describe the properties of metals since in a metal a part of the electrons is not located near the nuclei but are free to move in the volume of the metal sample. It is a surprisingly good approximation for some metals. It is a surprise since in the theoretical derivation of the properties of the gas; the electrons are seen as independent particles when in the real metals the free electrons are subject to their coulomb interaction. We shall not present explanation for the success of the model but just mention it.

In the second part of the chapter, we consider the case of semiconductors which are characterized by the absence of free electron when the temperature is zero. However, when the temperature increases, some electrons become free and form a gas of electron.

And the third case of a gas of fermions, we consider a star, a neutron stare making a gas of neutrons.

9.1 Free Electrons in a Box

9.1.1 *The Fermi–Dirac function*

We begin by the energy of the N fermions as a function of T, N and V. We recall the formula of the energy we got in Chapter 3:

$$E = \Sigma_i e_i \{\exp[\beta(e_i - \mu)] \pm 1\}^{-1} \tag{3.23}$$

In the present case, one takes the + sign and (3.23) becomes

$$E = \Sigma_i e_i \{\exp[\beta(e_i - \mu)] + 1\}^{-1} \tag{9.1}$$

or in passing to the continuum

$$E = \int_0^\infty E g(E) \{\exp[\beta(E_i - \mu)] + 1\}^{-1} dE \tag{9.2}$$

where $g(E)$ is the density of states for a massive particle with kinetic energy $p^2/2m$. The function $g(E)$ was already calculated in the preceding chapter and it is

$$g(E)dE = (8\pi V/h^3)\sqrt{2}m^{3/2}E^{1/2}dE \tag{9.3}$$

In (9.3) we introduced the factor 2 to take into account the spin of the electrons and neutrons (1/2). Thus the energy is given by

$$E = (8\pi V/h^3)\sqrt{2}m^{3/2} \int_0^\infty E^{3/2}\{\exp[\beta(E_i - \mu)] + 1\}^{-1} dE \tag{9.4}$$

In (9.2) and (9.3) the limits of the integrals are from 0 to ∞. One sees from (9.3) that, in principle, to calculate E one needs to calculate the chemical potential μ. For historical reason, it is frequently called the Fermi level. Sometimes the calling is restricted to the value of μ at $T = 0$ but in the present book, it will be called always "Fermi level" whatever the temperature.

The determination of μ is made through the general expression of N

$$N = \Sigma_i \{\exp[\beta(e_i - \mu)] \pm 1\}^{-1} \tag{3.22b}$$

or in our case

$$N = (8\pi V/h^3)\sqrt{2}m^{3/2}\int_0^\infty E^{1/2}\{\exp[\beta(E-\mu)]+1\}^{-1}dE \quad (9.5)$$

In order to be able to calculate μ from (9.4) and E from (9.3) a study of the function

$$f_{\mathrm{FD}} = \{\exp[\beta(E-\mu)]+1\}^{-1} \quad (9.6)$$

is useful. This function is called the Fermi–Dirac function. We recall that this function gives the probability that a state with energy E is occupied by an fermion.

First, we consider the FD function for $T = 0$ when the Fermi level is $\mu_0 > 0$. If the energy E is smaller than $\mu_0, \exp[\beta(E-\mu_0)]$ goes to 0 if $T \to 0(\beta \to \infty)$ since $E - \mu_0 < 0$. In this case f_{FD} goes to 1. Now if E is larger than $\mu_0, \exp[\beta(E-\mu_0)] \to \infty$ it $T \to 0(\beta \to \infty)$ since $E-\mu_0 > 0$. In this case, the function f_{FD} is null. Consequently the f_{FD} function is a step function passing from the value 1 to the value 0 when the variable E crosses the value $E = \mu_0$. The function is shown in Fig. 9.1.

We have already seen in Chapter 3 that the occupied levels are the N first levels. This means that at $T = 0$, the Fermi level is equal to the highest energy occupied by the last particle.

When the temperature increases (but it not too high), the f_{FD} function takes the shape indicated in Fig. 9.2 with $\mu > 0$ For $E \ll \mu$

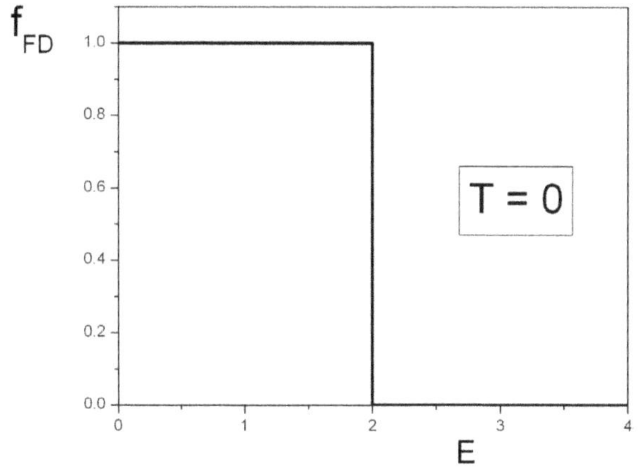

Fig. 9.1. The Fermi–Dirac function for $T = 0$. The energies are in electron–volt and the Fermi level is taken equal to 2 eV.

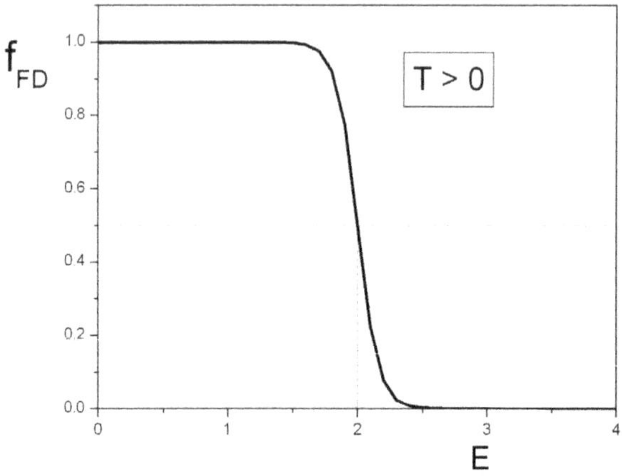

Fig. 9.2. The Fermi–Dirac function for $T > 0$. The energies are indicated in eV. In this figure, k_BT is equal to 0.08 eV. Since k_B is equal to 8.62 10^{-5} eV/°K, the temperature is 929°K.

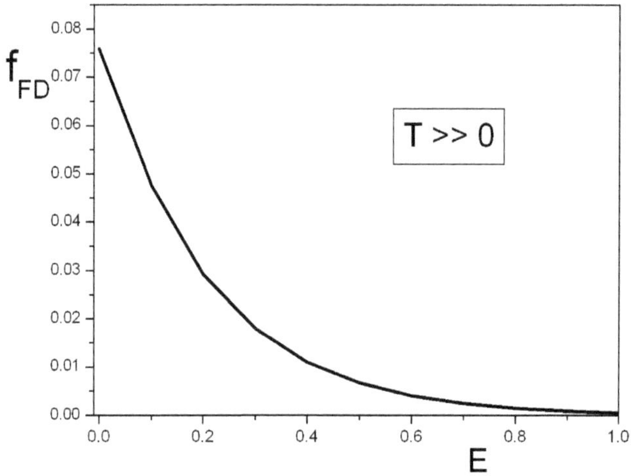

Fig. 9.3. The Fermi–Dirac function when the Fermi level is negative. It is equal to -0.5 eV and the temperature is equal to 2309°K. The energies are indicated in eV.

the function is equal to 1 but near μ, it decreases steeply to 0. For $E = \mu$ the function is equal to 1/2.

If now one increases strongly the temperature so that the Fermi level becomes negative, the f_{FD} can be approximate by an exponential: $f_{\text{FD}} = \exp[-\beta(E - \mu)]$ since $\exp[\beta(E - \mu)] \gg 1$ (see Fig. 9.3).

9.1.2 *The chemical potential or the Fermi level*

The variation of the Fermi level μ with the temperature can be calculated with the help of (9.5)

$$N = (8\pi V/h^3)\sqrt{2}m^{3/2}\int_0^{\infty} E^{1/2}\{\exp[\beta(E-\mu)]+1\}^{-1}dE \quad (9.5)$$

It is an implicit function of T and μ and it is clear that it is not possible to get an analytic expression giving $\mu(T)$. However at least for one value of T it is possible to get an exact value, i.e. for $T = 0$.

We saw in the preceding section that at $T = 0$, the Fermi–Dirac function is a step function equal to 1 for $E < \mu_0$ and equal to 0 for $E > \mu_0$. In such a case, the expression (9.5) reduces to

$$N = (8\pi V/h^3)\sqrt{2}m^{3/2}\int_0^{\mu} E^{1/2}dE \quad (9.7)$$

The integral is equal to $(2/3)(\mu_0)^{3.2}$. This gives for N

$$N = (8\pi V/h^3)\sqrt{2}m^{3/2}(2/3)\mu_0^{3.2} \quad (9.8)$$

From which one gets

$$\mu_0 = (N/V)^{2/3}(h^2/2m)(3/8\pi)^{2/3} \quad (9.9)$$

One takes a typical value of quantity for the free electrons in metals, $N/V = 5\ 10^{28}\,\mathrm{m}^{-3}$ and one finds for μ_0 a value of about 5 eV.

When the temperature is different from 0, we shall consider two different limits. For that, we transform the expression (9.5) using the trick already used several times. We multiply and divide the right-hand side of (9.4) by $(k_BT)^{3/2}$

$$\begin{aligned} N = {} & (8\pi V/h^3)\sqrt{2}m^{3/2}(k_BT)^{3/2} \\ & \times \int_0^{\infty} (E/k_BT)^{1/2} d(E/k_BT) \\ & \times \{\exp[(E-\mu)/k_BT)]+1\}^{-1} \end{aligned} \quad (9.10)$$

or

$$N/V = (4/\sqrt{\pi})[(2\pi m k_B T)/h^2]^{3/2} \times \int_0^\infty x^{1/2}[\exp(-\mu/k_B T)\exp(x) + 1]^{-1}dx \quad (9.11)$$

when we write $x = E/k_B T$. The two cases we consider are those with

$$(N/V) \ll (4/\sqrt{\pi})[(2\pi m k_B T)/h^2]^{3/2} \quad (9.12)$$

and

$$(N/V) \gg (4/\sqrt{\pi})[(2\pi m k_B T)/h^2]^{3/2} \quad (9.13)$$

We begin with the first condition (9.12) which means that the fermion density is small. When it is fulfilled, the integral of (9.11) (that we call I) reduces to

$$I = \int_0^\infty x^{1/2}[\exp(-\mu/k_B T)\exp(x) + 1]^{-1}dx \ll 1$$

This occurs when μ is negative and $\exp(-\mu k_B T) \gg 1$. In other words, the situation is classical or non-degenerate. In such a case the integral becomes

$$I = \exp(\mu/k_B T)\int_0^\infty x^{1/2}\exp(-x)dx = \exp(\mu/k_B T)(\sqrt{\pi}/2) \quad (9.14)$$

Inserting the value of the integral in (9.11) gives

$$(N/V) = 2[(2\pi m k_B T)/h^2]^{3/2}\exp(\mu/k_B T) \quad (9.15)$$

and one gets

$$\mu = k_B T \text{ Ln}(N/V) - k_B T \text{ Ln}\{2[(2\pi m k_B T)/h^2]^{3/2}\} \quad (9.16)$$

We have already found all these results. The inequality (9.12) is the condition for the classical case and is equivalent to the condition (4.16). And the expression (9.16) for the chemical potential is the equivalent of the expression (3.29) when one inserts the complete expression (4.7) for Z_1. The only difference stands in a factor 2 appearing in (9.16) since we took into account the spin.

Now we consider the case given by the inequality (9.13) which corresponds to a large fermion density and relatively low temperatures. It is the degenerate case. One needs to use the complete expressions (9.4) or (9.5). As mentioned above, it is not possible to get an analytic expression $\mu(T)$ giving the variation of the Fermi level with the temperature. But a numerical calculation is feasible and we performed it in order to give an illustration of the variations of μ with T.

We chose $\mu_0 = 2\,\text{eV}$ which corresponds to a electron density of about $5\,10^{26}\,\text{m}^{-3}$. We verify that the inequality (9.13) in calculating the quantity $(4/\sqrt{\pi})[(2\pi m k_B T)/h^2]^{3/2}$ for $= 500°\text{K}$. We get $3.4\,10^{25}\text{m}^{-3}$, i.e. the inequality (9.13) is just satisfied. In such case the complete curve (Fig. 9.4) shows that the Fermi level is practically constant up to this temperature. Important changes in μ appear only at very high temperatures. μ decreases until it becomes null for $T = 33540°\text{K}$. Of course no element can stand this temperature and we show the complete curve for the sake of the completeness. The conclusion is that for large densities and "reasonable" temperatures the Fermi level is constant,

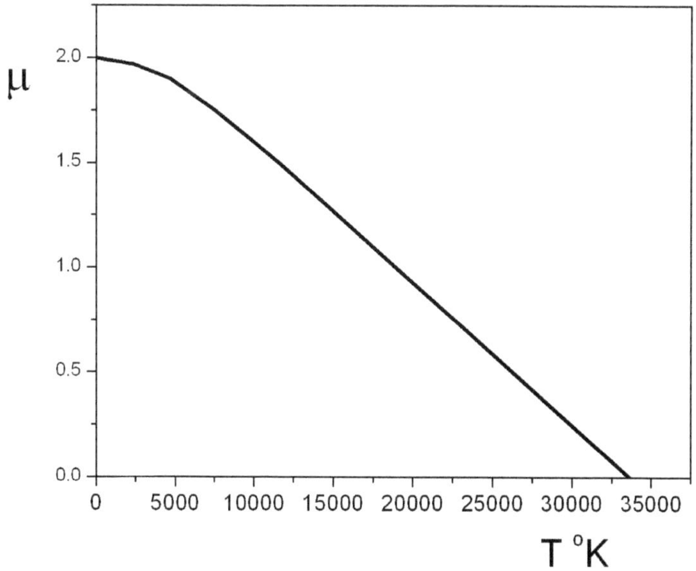

Fig. 9.4. Variation of the Fermi level with the temperature.

We can now confess that it is possible to have (after hard calculation work) an expression for the Fermi level at low temperature. It is

$$\mu = \mu_0 - (\pi^2/12)(k_B T)^2/\mu_0 \tag{9.17}$$

The reader can verify (for example, for $T = 500°\mathrm{K}$) that the relative change in the Fermi level $(\mu - \mu_0)/\mu_0$ is really very small such that the conclusion of the preceding paragraph is correct.

9.1.3 *The energy*

(A) $T = 0$

The calculation of the energy E at $T = 0$ is not difficult due to the simple form of the Fermi–Dirac function. We recall the general expression for E (9.4)

$$E = (8\pi V/h^3)\sqrt{2}m^{3/2} \int_0^\infty E^{3/2}\{\exp[\beta(E_i - \mu)] + 1\}^{-1} dE \tag{9.4}$$

The expression of E becomes

$$E(0) = (8\pi V/h^3)\sqrt{2}m^{3/2} \int_0^\mu E^{3/2} dE \tag{9.18}$$

The integral is equal to $(2/5)\mu_0^{5/2}$ and one has

$$E(0) = (8\pi V/h^3)\sqrt{2}m^{3/2}(2/5)\mu_0^{5/2} \tag{9.19}$$

We recall a preceding expression

$$N = (8\pi V/h^3)\sqrt{2}m^{3/2}(2/3)\mu 0^{3.2} \tag{9.8}$$

And we divide side by side the expression (9.19) and (9.8): to get

$$E(0)/N = (3/5)\mu_0 = 0.0727(N/V)^{2/3}(h^2/2m) \tag{9.20}$$

(B) $T \neq 0$ The first method

The determination of E for $T \neq 0$ is much more difficult and before we derive an expression valid at low temperatures, we need to recall two properties relative to definite integrals, which are not well known

Property 1. The permutation of the limits of the integral changes its sign

$$\int_a^b f(x)dx = -\int_b^a f(x)dx$$

Property 2. It is applicable if the function under the integral is defined for positive and negative values of the variable

$$\int_{-a}^0 f(x)dx = \int_0^a f(-x)dx$$

We begin with the expression (9.4) that we write in the following form:

$$E = D\int_0^\infty h(E)\{\exp[(E-\mu)/k_BT]+1\}^{-1}dE \qquad (9.21)$$

where $D = (8\pi V/h^3)\sqrt{2}m^{3/2}$ and $h(E) = Eg(E) = E^{3/2}$.

The Fermi level appears in (9.21) and in principle it is temperature dependent. But in the temperature range we consider we take it constant and equal to its value at $T = 0$ as explained in the preceding section.

We have to calculate the integral.

$$J = \int_0^\infty h(E)\{\exp[(E-\mu)/k_BT]+1\}^{-1}dE \qquad (9.22)$$

in the limit of low temperatures. As a first step one makes a change in the variable $x = (E-\mu)/k_BT$ and one has $E = \mu + k_BTx, dE = k_BTdx$ and the limits are now $-\mu/(k_BT)$ and ∞.

The integral J becomes

$$J = k_BT\int_{-\mu/kT}^\infty h(\mu + k_BTx)[\exp x + 1]^{-1}dx \qquad (9.23)$$

In the second step, the integral is written as the sum of two integrals

$$J = J_1 + J_2$$

$$J_1 = k_BT\int_{-\mu/kT}^0 h(\mu + k_BTx)[\exp(x) + 1]^{-1}dx \qquad (9.24a)$$

$$J_2 = k_BT\int_0^\infty h(\mu + k_BTx)[\exp(x) + 1]^{-1}dx \qquad (9.24b)$$

Using the property 2 above the integral J_1 becomes

$$J_1 = k_BT \int_0^{\mu/kT} h(\mu - k_BTx)[\exp(-x) + 1]^{-1}dx \tag{9.25}$$

Using the identity

$$[\exp(-x) + 1]^{-1} = 1 - [\exp(x) + 1]^{-1}$$

and putting it in (9.25), one sees that J_1 can be written as the sum of two integrals J_1' and J_1'':

$$J_1' = k_BT \int_0^{\mu/kT} h(\mu - k_BTx)dx \tag{9.26}$$

$$J_1'' = -k_BT \int_0^{\mu/kT} h(\mu - k_BTx)[\exp(x) + 1]^{-1}dx \tag{9.27}$$

Now we make in J_1' the following change of the variable x: $y = \mu - k_BTx, dx = -dy/(k_BT)$ and the limits become $y = \mu$ and $y = 0$. Consequently

$$J_1' = -k_BT \int_\mu^0 h(y)dy/(k_BT) = \int_0^\mu h(y)dy \tag{9.28}$$

The second equality is due to the property 1 above.

The upper limit of the integral J_1'' is $\mu/(k_BT)$ which is very large when T is very small thus we take it infinite. We can give to the integrals J_2 and J_1'' the same limits $[0, \infty]$ and put them together in one sum

$$J''' = k_BT \int_0^\infty [h(\mu + k_BTx) - h(\mu - k_BTx)][\exp x + 1]^{-1}dx \tag{9.29}$$

It is an important result since $J = J_1' + J'''$ and it can be used if the function $h(y)$ has a simple form like a polynom. In such a case it is possible to calculate the integral since the values integrals of the type $\int_0^\infty x^n[\exp(x)+1]^{-1}$ are known (n is an integer). But in the present case $h(y) = y^{3/2}$ and the integral (9.29) cannot be calculated.

We shall develop the function $h(y)$ appearing in the integrand of J''' in series around the value μ. It is justified as long as $(k_BTx) = \varepsilon$

is much smaller than μ and when it is not the case (since x goes to infinite), $\exp(x)$ becomes large enough to make the integrand small and negligible.

One has

$$h(\mu-\varepsilon) = h(\mu) - \varepsilon(dh/dy)|_\mu + \varepsilon^2(d^2h/dy^2)_\mu/2 + \cdots$$
$$h(\mu+\varepsilon) = h(\mu) + \varepsilon(dh/dy)|_\mu + \varepsilon^2(d^2h/dy^2)_\mu/2 +$$

and

$$h(\mu+\varepsilon) - h(\mu-\varepsilon) = \varepsilon(dh/dy)_\mu$$

Finally the integral J''' is

$$J''' = (k_BT)^2(dh/dy)_\mu \int_0^\infty x[\exp x + 1]^{-1}dx \qquad (9.30)$$

The integral $\int_0^\infty x[\exp x + 1]^{-1}dx$ is equal to $\pi^2/6$.
The final result is

$$J = J_1 + J_2 = J_1' + J_1'' + J_2$$
$$J = J_1' + J'''$$

Putting the value of J_1' (9.28) and that of J''' (9.30) gives

$$J = \int_0^\mu h(y)dy + \pi^2/6(k_BT)^2(dh/dy)_\mu \qquad (9.31)$$

We come back to the initial expression of the energy and introduce the integral J in (9.21)

$$E = DJ = (8\pi V/h^3)\sqrt{2}m^{3/2}$$
$$\times \left[\int_0^\mu h(y)dy + \pi^2/6(k_BT)^2(dh/dy)|_\mu\right] \qquad (9.32)$$

Recalling that $h(y) = y^{3/2}$ and $(dh/dy)|_\mu = 3/2\,\mu^{1/2}$

$$E = (8\pi V/h^3)\sqrt{2}m^{3/2}\int_0^\mu y^{3/2}dy + (8\pi V/h^3)$$
$$\times \sqrt{2}m^{3/2}\pi^2/6(k_BT)^2(3/2)\mu^{1/2} \qquad (9.33)$$

or

$$E = E(0) + (8\pi V/h^3)\sqrt{2}m^{3/2}\pi^2/6(k_BT)^2(3/2)\mu^{1/2}$$

The first term is the energy for $T = 0$ (see expression (9.18)) when the integrand y replaces E and the second term is the contribution due to the temperature.

(C) The second method

This is a method inspired from the work of Sommerfeld and consists in a different calculation of the integral J above. We recall that $h(E)$ is the energy E multiplied by $G(E)$, the density of states

$$J = \int_0^\infty h(E)\{\exp[(E-\mu)/k_BT]+1\}^{-1}dE \qquad (9.22)$$

The first step is a change in the variable and the new variable x is given by $(E-\mu) = k_BTx$. Thus

$$J = \int_{-\frac{\mu}{\mathrm{kT}}}^\infty dxh(\mu + k_BTx)/(\exp(x)+1)$$

The lower limit of the integral is now $a = -\mu/k_BT$. Since we intend to make this calculation at very low T, it is possible to take $a = -\infty$ and one has

$$J = \frac{\int_{-\infty}^\infty dxh(\mu + k_BTx)}{\exp(x)+1}$$

The second step is to transform integral J by integration by parts. One has

$$du = h = (\mu + k_BTx)^{\frac{3}{2}} \quad \text{and} \quad u = 2/(5k_BT)(\mu + k_BTx)^{5/2}$$
$$v = 1/(\exp(x)+1) \quad \text{and} \quad dv = (xk_BT)(\exp(x)/(\exp(x)+1)2$$

The integral J becomes

$$J = \int_{-\infty}^\infty \left(\frac{2}{5}\right)(\mu + k_BTx)^{5/2}(x\exp(x))/(\exp(x)+1)^2$$

The next step is to develop $(2/5)(\mu + k_BTx)^{5/2}$ around x which is a small quantity. Following the schema

$$\begin{aligned} f(x) &= f(0) + xf'(0) + x^2/2f'''(0) \\ &= (2/5)(\mu + k_BTx)^{5/2} \\ &= (2/5)\mu^{5/2} + (k_BT)\mu^{3/2}x + (3/2)(k_BT)^2\mu^{1/2}x^2/2 \end{aligned}$$

The integral J is now divided in three integrals. The first is

$$\int_{-\infty}^{\infty} \left(\frac{2}{5}\right) (\mu)^{5/2}(\exp(x))/(\exp(x)+1)^2 dx/2$$

and it is equal to $2/5(\mu)^{5/2}$.

The second integral is

$$\int_{-\infty}^{\infty} \mu^{\frac{3}{2}}(\mathrm{dx})x\frac{kT\exp(x)}{(\exp(x)+1)^2} = 0$$

since the integrand is symmetric relatively to the x and finally the third integral is

$$\int_{-\infty}^{\infty} \left(\frac{3}{4}\right) \mu^{\frac{1}{2}}(k_BT)^2 \frac{\exp(x)x^2}{(\exp(x)+1)^2} dx = (3/2)\mu^{1/2}(k_BT)^2(\pi^2/6)$$

The final result is identical to that of the first method as the reader can verify. When comparing two methods, it is clear that the first is more complicated. However its final formula is easier to be used in several cases We recall the formula

$$J = \int_0^{\mu} h(y)dy + \pi^2/6(k_BT)^2(dh/dy)_\mu \tag{9.31}$$

in which it is only necessary to know the function $h(y)$ which can be known for various situations (see exercises).

(D) Résumé of the first method

The goal is to calculate the integral of their expression

$$h(E)\{\exp[(E+\mu)/k_BT]+1\}^{-1}dE$$

where $h(E)$ is the density of states multiplied by the energy.

After a change in the variables, the integral is divided into two integrals

$$J_1 = k_BT \int_{-\mu/kT}^{0} h(\mu + k_BTx)[\exp(x) + 1]^{-1}dx \qquad (9.24a)$$

$$J_2 = k_BT \int_{0}^{\infty} h(\mu + k_BTx)[\exp(x) + 1]^{-1}dx \qquad (9.24b)$$

After some manipulations, it is possible to write J as

$$J = \int_0^\infty h(y)dy + k_BT \int_0^\infty [h(\mu + kTx) - h(\mu - kTx)]dx/(\exp(x) + 1)$$

The first integral gives $E(0)$ and the second the energy for $T > 0$.

$$E(T) = J''' = (k_BT)^2(dh/dy)_\mu \int_0^\infty x[\exp(x) + 1]^{-1}$$

(E) The formulas of the energy at $T = 0$ and at $T \neq 0$

Energy at all temperatures

$$E = (PV)/K \quad \text{and} \quad K = 3/2$$

Energy at $T = 0$
$E(0) = 3/5N_{\mu 0}$

$$E(0) = (8\pi V/h^3)\sqrt{2}m^{3/2}(2/5)\mu_0^{5/2}$$

Energy at $T \neq 0$

$$E = E(0) + (8\pi V/h^3)\sqrt{2}m^{3/2}\pi^2/6(k_BT)^2(3/2)\mu^{1/2}$$

The quantity $(8\pi V/h^3)\sqrt{2}m^{3/2}$ appears in the relation between N and μ_0

$$N = (8\pi V/h^3)\sqrt{2}m^{3/2}(2/3)\mu_0^{3.2} \qquad (9.8)$$

And one deduces that $(8\pi V/h^3)\sqrt{2}m^{3/2} = (3N/2\mu_0^{3.2})$. Introducing it in (9.31) gives the final expression[1] for E at low temperatures:

$$E = E(0) + (\pi^2/4)N(k_BT)^2\mu_0 \qquad (9.34a)$$

[1]In several textbooks the expression for the energy is derived with the help of the method due to Sommerfeld. The first method used here is due to Landau and Lifshitz.

with $E(0) = (3/5)N\mu_0$. Introducing the value of μ_0 in (9.34a) gives

$$E = E(0) + 10.17N(V/N)^{2/3}(k_BT)^2(2m/h^2) \tag{9.34b}$$

The specific heat

From the expression of the energy (9.34a) one can calculate the specific heat at constant volume $C_V = (\partial E/\partial T)_V$

$$C_V = (\pi^2/2)N(k_B)^2T/\mu_0 \tag{9.35}$$

The specific heat is linear with the temperature.

In several metals this relation is verified but at very low temperature. The total specific heat of a metal is compound of the contribution of the atoms and that of the free electrons. However the first contribution is much larger than the second except at low temperature when the atomic contribution is also very low because it varies with the temperature as T^3.

It is possible to write the specific heat in different forms. One can introduce the Fermi temperature T_F defined by $k_BT_F = \mu_0$ or $T_F = \mu_0/k_B$. Putting in (9.33) gives

$$C_V = (\pi^2/2)k_BN(T/T_F) \tag{9.36}$$

Finally we give another expression for C_V. We recall the expression of the number of electrons

$$N = (8\pi V/h^3)\sqrt{2}m^{3/2}\int_0^\infty E^{1/2}\{\exp[\beta(E-\mu)]+1\}^{-1}dE \tag{9.5}$$

which can be written as $N = \int_0^\infty n(E)dE$. The function $n(E)$ can be interpreted as giving the number of electrons in the vicinity of the energy E. For $E = \mu_0$ one has

$$n(\mu_0) = (8\pi V/h^3)\sqrt{2}m^{3/2}\mu_0^{1/2} \tag{9.37}$$

This gives

$$\mu_0^{1/2} = n(\mu_0)[(8\pi V/h^3)\sqrt{2}m^{3/2}]^{-1} \tag{9.38}$$

Above we found that $(8\pi V/h^3)\sqrt{2}m^{3/2} = (3N/2\mu_0^{3.2})\cdot$ and putting it in (9.38) gives $n(\mu_0) = (3N/2\mu_0)$ or $\mu_0 = (3N/2[n(\mu_0)]$. Inserting this expression of μ_0 in (9.34a)

$$E = E(0) + \pi^2/6(k_BT)^2 n(\mu_0) \tag{9.39}$$

One gets for C_V

$$C_V = \pi^2/3n(\mu_0)k_B^2 T \tag{9.40}$$

The interest of this last formula stands in the possibility to extract $n(\mu_0)$, the number of electrons with energy equal to the Fermi level, from a measure of the electronic specific heat.

9.2 Applications to Metals

We shall give a simplified picture of the energies of electrons in solids and in particular in metals. The energy levels of electrons in an isolated atom are a series of discrete values but when the atoms form a solid, the possible energies for electrons are grouped in bands. The origin of the bands is the fact that the atoms are not isolated but interact in order to form a solid. It results that the density of states for electrons in solids is compound of several distinct regions called "bands". In Figs. 9.5 and 9.6 we give the two possible situations for

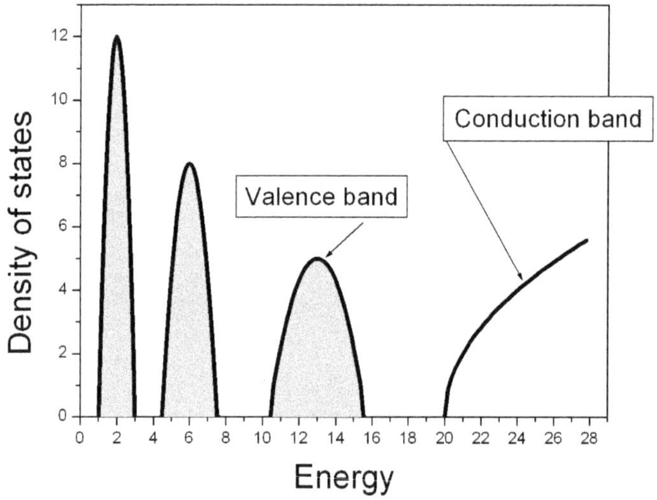

Fig. 9.5. Schematic picture of the electronic bands in an insulator.

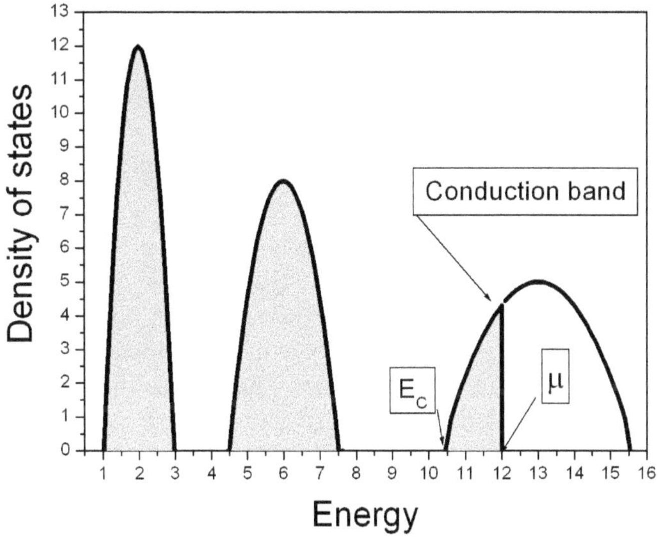

Fig. 9.6. Schematic picture of the electronic bands in a metal.

solids at temperature zero. In the first (Fig. 9.5) the first bands are completely occupied by the electrons and the last band is empty. In Fig. 9.5, we show an example of three occupied bands and one empty band. It is possible to show that a filled band cannot sustain an electric current and thus the material is an insulator.

In the other situation, the last band is not completely filled. We show in Fig. 9.6 such an example of two bands completely filled when the third is only partially filled. This band is called the conduction band since the electrons in this band can sustain an electric current. This material is a metal. The electrons of the conduction band are free electrons which are not bonded to specific atoms contrarily to the case of insulators.

In a first approximation, the picture of free electrons we gave above can be applied to the electrons in the conduction band of metals but with two changes. In the above calculations it was supposed that the smallest energy is zero when in a metal, the smallest energy is that of the bottom E_C of the conduction band. Thus each time that the energy E appears in the above expressions, we have to replace it by $E - E_c$. However this does not modify the expressions of the specific heat. The second change concerns the mass of the electron.

One of the influences of the atoms on the free electrons is to change their mass which is now an effective mass. But this effect is not strong such that in our approximation it is possible to use the electron mass.

9.3 Electrons in Semiconductors

Semiconductors are a particular class of insulators. Like all insulators, the electrons are bonded to the nucleus such that at zero temperature there are no free electrons. But it is possible to liberate electrons from the atoms and consequently the insulator can sustain an electric current. In the band picture this means that electrons with energy of the last filled band, called the valence band, have gained enough energy to pass from the valence band to the conduction band. In semiconductors this process takes place by heating the material (and also under influence of the light).

Once the electrons are liberated and have their energy in the conduction band, what we said above about the free electrons in metals is applicable concerning their energy minimum (E_c) and their effective mass. But there is an important difference: the Fermi level is below the minimum energy E_C of the electrons when for metals it is in the conduction band. Consequently the electrons do not obey the Fermi–Dirac statistics but the Maxwell–Boltzmann statistics. This can be understood in considering Fig. 9.5. Since at $T = 0$ the conduction band is empty and the valence band filled, from the property of the Fermi–Dirac function, one sees that the Fermi level must be between E_C, the bottom of the conduction band and E_V the top of the valence band: $E_V < \mu < E_c$. We want to answer the following question: what is the number of electrons in the conduction band when the material is heated? It is important in order to understand the electric conduction in the semiconductors which depends strongly on the temperature. It is not the case of the electrons in metals since their number does not depend on the temperature.[2]

[2]It does not mean that their electric conduction is completely independent of the temperature since the electric conduction is a function of the number of free electrons but also of their ability to move under influence of the electric field. This last property is slightly dependent on T.

To calculate the number of electrons in the conduction band, one needs to know the density of states. For that we use the "parabolic approximation". We assume that the conduction band can be approximate, in the vicinity of E_C, by the following expression:

$$g_c(E) = D_C(E - E_C)^{1/2} \tag{9.41}$$

in analogy with the case of free electrons in a box (see (9.3)). We complete the analogy in writing $D_C = (8\pi V/h^3)\sqrt{2}m_e^{3/2}\, m$ where m_e is the effective mass of an electrons (see Fig. 9.7). The effective electron mass is defined through D_C and it is often smaller than the electron mass.

The number of electrons in the conduction band is now

$$N_e = (8\pi V/h^3)\sqrt{2}m_e^{3/2} \int_{E_C}^{\infty} (E - E_C)^{1/2}\{\exp[\beta(E - \mu)] + 1\}^{-1} dE$$

The limits of the integral are chosen from E_C to ∞. The lower limit is self evident but the upper one is not correct since the parabolic approximation is valid only for values of E near E_C. However the Fermi–Dirac function is a rapidly decreasing function such that for large values of E the integrand is very small and does not

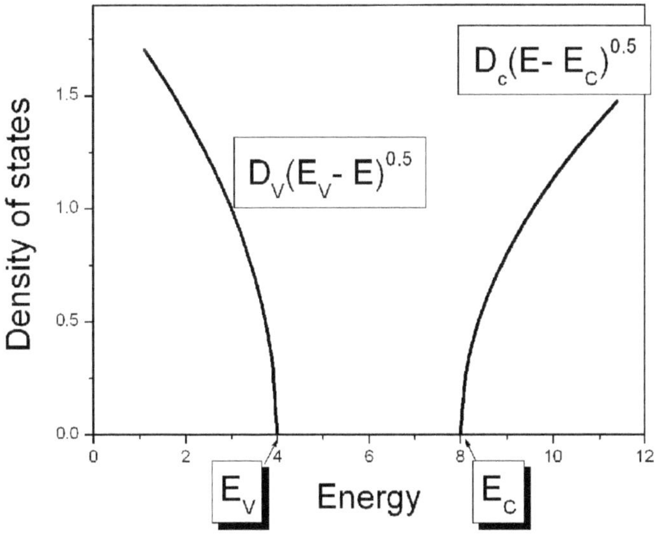

Fig. 9.7. Parabolic approximation of the bands of a semiconductor.

contribute significantly to the integral. Since the Fermi level μ is smaller than E_C the Fermi–Dirac function reduces to an exponential function and one has

$$N_e = (8\pi V/h^3)\sqrt{2}m_e^{3/2} \int_{E_C}^{\infty} (E - E_C)^{1/2} \exp[-\beta(E - \mu)]dE \quad (9.42)$$

Now we deal with the electrons in the valence band. In this band there is a lack of electrons equal to the number of electrons in the conduction band. Instead to consider the electrons themselves in the valence band, we consider the number of states which are not occupied by the electrons. If the probability for a state to be occupied by an electron is given by the Fermi–Dirac function, the probability that a state is not occupied is $1 - f_{\text{FD}}$. The number of unoccupied states is equal to the density of states in the valence band multiplied by $(1 - f_{\text{FD}})$.

The unoccupied states of the valence band can receive an intuitive picture. When an electron is bonded to an atom, it moves in a restricted region of the space near the atom. If now this electron is liberated and moves freely, this unoccupied place may receive one electron from the neighboring atoms. In leaving the atom this new electron creates a new place which can be filled by another electron. Thus the appearance of a free electron brings about the motion of electrons from unoccupied place to another. These wandering places correspond to the unoccupied states of the valence bands. There are called "holes".

We use again the "parabolic approximation" for the valence band as shown in Fig. 9.7 and write

$$g_v(E) = D_v(E_V - E)^{1/2} \quad (9.43)$$

and $D_v = (8\pi V/h^3)\sqrt{2}m_h^{3/2}$ where one defines m_h as the effective mass of a hole.

The number of the holes is given by

$$N_V = (8\pi V/h^3)\sqrt{2}m_h^{3/2} \int_{-\infty}^{E_V} (E_V - E)^{1/2}(1 - f_{\text{FD}})dE \quad (9.44)$$

where

$$1 - f_{\text{FD}} = 1 - \{\exp[\beta(E - \mu)] + 1\}^{-1} = \{\exp[-\beta(\mu - E)] + 1\}^{-1}$$

Since $E < \mu$ one has $\exp[-\beta(E-\mu)] \gg 1$ and $1 - f_{\rm FD}$ becomes $\exp[\beta(\mu - E)]$. Inserting in (9.44) gives

$$N_V = (8\pi V/h^3)\sqrt{2}m_h^{3/2}\int_{-\infty}^{E_V}(E_V - E)^{1/2}\exp[-\beta(\mu - E)]\,dE \quad (9.45)$$

Now we are able to calculate the Fermi level as function of the temperature and the number of electrons in the conduction band which is equal to the number of holes in the valence band. For that we shall make some transformations to the expressions (9.42) and (9.45).

First we write in (9.42), $E - \mu = E - E_C - (\mu - E_C)$

$$\beta(E - E_C) = (E - E_C)/(k_BT) = y\,dE = (k_BT)dy$$

The expression (9.42) is now

$$N_e = (8\pi V/h^3)\sqrt{2}(k_BT)^{3/2}m_e^{3/2}\exp[\beta(\mu - E_C)] \times \int_0^{\infty} y^{1/2}\exp(-y)dy \quad (9.46)$$

In (9.45) we write

$$\mu - E = \mu - E_V + (E_V - E)$$
$$\beta(E_V - E) = (E_V - E)/(k_BT) = y\,dE = -(k_BT)/dy$$

and

$$N_V = (8\pi V/h^3)\sqrt{2}(k_BT)^{3/2}m_h^{3/2}\exp[-\beta(\mu - E_V)] \times \int_0^{\infty} y^{1/2}\exp(-y)dy \quad (9.47)$$

In writing (9.47) we used the property 1 above concerning definite integrals. Now we have two equations to find the two unknown quantities: $N_C = N_V$ and μ. The value of the integral appearing in (9.46) and (9.47) is $\sqrt{\pi}/2$. First, we multiply (9.46) and (9.47) side by side and one finds

$$N_CN_V = [(32\pi^3V^2)/h^6](k_BT)^3(m_em_h)^{3/2}\exp[-\beta(E_C - E_V)]$$

and finally

$$N_C = N_V = (N_C N_V)^{1/2} = 2V[(2\pi k_B T)/h^2]^{3/2}(m_e m_h)^{3/4} \times \exp[-\beta(E_C - E_V)/2] \tag{9.48}$$

One sees that the number of electrons in the conduction band and of the holes in the valence band is strongly dependent on the temperature. In particular it is a function of the difference $E_C - E_V$ which is called the band gap.

To get the Fermi level, one writes that the right-hand sides of (9.46) and (9.47) are equal. One obtains

$$\mu = (E_C + E_V)/2 + (3/4)k_B T\ \mathrm{Ln}(m_h/m_e) \tag{9.49}$$

For $T = 0$ the Fermi level is exactly at mid distance from the bottom of the conduction band and the top of the valence band. Since frequently $m_h > m_e$, the Fermi level is an increasing function of T.

All what we describe about the properties of the semiconductor concerns the case where there are not impurities. These impurities may liberate some of their electrons in increasing the temperature. They introduce electrons and holes. When their numbers are higher than the electrons and holes from the material itself, one speaks about the extrinsic regime.

A final word about the holes as missing places of electrons. In the absence of electric field, the jumping electrons and the missing places are moving at random. But under influence of an electric field, the electrons move preferentially in the opposite direction of the field. Consequently the missing places move in the direction of the field. The missing places behave as if there were particles with a positive electric charge and the effective mass of the holes.

9.4 Neutron Stars

We shall terminate the chapter by another example of a fermion gas. They are stars which are made of neutrons. These stars are created by evolution of larger stars and may disappear after some time as black holes. They have specific properties, in particular their huge density. The distance between two neutrons is of the same order of magnitude than that of nucleons in the atoms. The statistical mechanics will help us to get an evaluation of the typical size of a neutron star.

The neutron stars have their stability from two elements. The first is the kinetic energy E_K of the fermion gas and the second is the gravitational energy. Thus the two kinds of energy will have opposite contributions to the total energy We shall first calculate this two energies and afterward determine what is the volume which gives the minimum of the total energy.

9.4.1 *The kinetic energy of the neutrons*

The determination of the energy of the neutrons stands on several assumptions. The first is that the neutron stars are degenerated fermions gas. One supposes also that the gas is nonrelativist. And finally the calculation is made at temperature zero.

The energy at $T = 0$ is given by

$$E(0) = 3/5\, N\mu_0 \quad \text{and one has } \mu_0 = (N/V)^{2/3}(h^2/2m)(3/8\pi)^{2/3}$$

This gives

$$E_K = 0.073(N^{5/3}/V^{2/3})(h^2/2m) \tag{9.50}$$

One recall that is m the mass of a neutron

9.4.2 *The gravitational energy of a sphere*

One begins by the potential of a system of two masses, M and m: $U = -GMm/R$. Now consider a mass in form of a sphere with radius R. The density ρ is uniform such that the mass of sphere is

$$M = \rho V$$

One delimits inside the mass M a sphere with radius r smaller than R and a layer of thickness dr around the sphere r. dr must be understood as the differential of r. The mass of this layer is

$$m = 4\pi r^2 \rho dr$$

Applying the formula above to the mass M which is the mass of the sphere with radius R and the mass m that of the layer. One has the differential of the energy U as

$$dU = -(4/3)\pi G r^3 \rho (4\pi r^2 \rho dr)/r$$

and the energy itself as

$$U = -\int_0^R (4/3)\pi G \rho r^3) \frac{4\pi \rho r^2 dr}{r} = -(16/3)\pi^2 \rho^2 R^5/5$$

where

$$\int_0^R r^4 dr = \frac{R^5}{5}$$

$$\mathrm{U} = -G[(4/3)(\pi^2 \rho^2 R^3)]^2 (3/5)(1/R)$$

And finally

$$U = -\left(\frac{3}{5}\right)\frac{GM^2}{R} \tag{9.51}$$

The total energy is

$$E = E_K + U = 3/5\ N\mu_0 - \left(\frac{3}{5}\right)\frac{GM^2}{R} \tag{9.52}$$

9.4.3 *The volume and the radius of a neutron star*

One has to minimize the energy relatively to the volume or the radius. It is also necessary to work with either the mass or the number N of particles. One choses the radius and the mass. $M = Nm$. The energy is written as

$$E = \left(\frac{3}{5}\right)\left(N\mu_0 - \frac{GM^2}{R}\right) \tag{9.53}$$

And the derivative relatively to R is

$$\frac{\partial E}{\partial R} = -(3/5)\left[N(\partial\mu_0/\partial V)\left(\frac{\partial V}{\partial R}\right) + GM^2R^{-2}\right] = 0$$

$$\partial\mu_0/\partial V = -(2/3)V^{-1/3}(N)^{5/3}(h^2/2m)(3/8\pi)^{2/3}$$

Writing $N = M/m$, one gets the equation for which the energy is minimum

$$R = (3/8\pi)^{2/3}(3/4\pi)^{1/3}[h^2/(GM^{1/3}m^{8/3})] \tag{9.54}$$

Now one makes a numeral calculation of radius. The constant G is equal to $6.7\ 10^{-11}$ (in MKS) and the mass of a neutron is $1.7\ 10^{-27}$ kg. The mass of a neutron star is of order of magnitude of the mass of the sun $M = 3\ 10^{30}$ kg. (the mass of some neutron stars has been determined, here we take 1.5 time the mass sun) This gives for $R \approx 10$ km which is very small size comparatively to other stars like the sun that has a radius of $7\ 10^5$ km.

Chapter 10

A History of Statistical Mechanics

10.1 Thermodynamics and Statistical Mechanics Before Maxwell and Boltzmann

It is impossible to write a history of statistical mechanics without reference to thermodynamics. These fields are two aspects of the more general field of thermal physics which includes the macroscopic and macroscopic sides. However, I shall try to emphasize the important steps which permitted an understanding of the development of statistical mechanics. It is also difficult in a limited space of one chapter (a complete book is necessary) to quote all the scientists who contributed to the advances in the field.

The first precursor of statistical mechanics is probably the mathematician Daniel Bernoulli who wrote a treatise of hydrodynamics in 1738. He gave a picture of a gas made of balls bouncing on the walls of the container. He calculated the pressure on the walls by the motion of the ball and he derived the Boyle law of the constancy of the product (pressure)(volume) at constant temperature. He showed also that this product is proportional to the kinetic energy of the balls. One has to note that it is supposed that all the balls have the same velocity. This assumption is adopted by numerous scientists and the work of Maxwell described below departs from this view. The work of Bernoulli was forgotten until 1859.

Thermodynamics begins really at middle of the eightieth century when the properties of gas began to be known (Boyle law, Gay Lussac law) (at this time the concept of ideal gas was not known and it was thought that all the gas have the same properties). The use of the heat to produce work began to be realized in steam engines. The concepts of specific heats and of latent heat are developed. Like in all fields of science experimental works and theoretical works appeared side by side. At the beginning of the ninetieth century there is a new area with the formulation of the important concept of energy (Young in 1806), the study the heat propagation (Fourier in 1807), thermal properties of solids (Dulong and Petit in 1819) and of gas (Delaroch et Bérard in 1812) and the problem of understanding the mechanism of the heat engine in order to maximize the work obtained from heat. The work of Sadi Carnot in 1824 concerning the transformation of heat in work is fundamental since he gave the basic principles of a cycling engine. He showed that a fraction of the heat that the engine receives from the heat reservoir is expelled outside without being transformed in work.

In the middle of the nineteenth century the basis of thermodynamics was established with the two laws of thermodynamics. In particular the second law received several formulations. One of the most important formulations is based on the concept of entropy introduced by Clausius[1] in 1865. Formally it is proposed that exists a state function of a system defined by the integral taken between two equilibrium states $S = \int dQ/T$ (dQ is a reversible small amount of heat) It was a mysterious concept related to irreversibility. It was first related to the lack of symmetry concerning the transformation of work in heat and the contrary. It is possible to transform some work completely in heat the converse in not true as shown by Carnot. But the most impressive aspect is the principle that in an isolated system any irreversible process (or any process taking place spontaneously) results in an increase of the entropy.

In parallel of the development of thermodynamics several theoretical works were done on the theory of gas by Herapath in 1812, Laplace in 1824 on the interaction between particles, Waterston in 1845, Joule in 1851, Clausius in 1857. But they were not in the mean stream of interest for majority of scientists. One important result of these works is that the mean kinetic energy of one

[1]It is usual to mention the entropy by the letter S. It was the letter used for the first time by Clausius and it was speculated that maybe it was done in honor of Sadi Carnot.

molecule of a gas is proportional to the temperature or more exactly $1/2(m\mathrm{v}_m{}^2) = (3/2)(R/N_A)(273 + T)(N_A$ is Avogadro number, R is the gas constant and T is given in centigrade degrees).

10.2 The Kinetic Theory of Maxwell

The name of Maxwell is well known by the fundamental equations of electromagnetism but he worked also in the kinetic theory of gas. In 1860 he published the determination of the velocity distribution of an ideal gas.

The goal of this work is to find what proportion of N molecules in a gas has a velocity with values between v and $v + dv$. One begins by the unknown function $f(v_x)$ which gives the number of molecules with velocity in the x direction between v_x and $v_x + dv_x$. This number is $Nf(v_x)dv_x$. In other direction y and z, one has similar expressions for the number of molecules with velocity between v_y and $v_y + dv_y \cdot Nf(v_y)dv_y$ and in the z direction $Nf(v_z)dv_z$ since the gas is isotropic. Now for the number of molecules with velocities between v_x and $v_x + dv_x$, v_y and $v_y + dv_y \cdot v_y$ and $v_z + dv_z$ one has $Nf(v_x)f(v_y)f(v_z)dv_x\, dv_x dv_z$ since the velocities in the different directions are independent. But since there is no correlation between the velocities, this number of molecules must depend only the total velocity $(v_x^2 + v_y^2 + v_z^2)^{1/2}$ or

$$f(v_x)f(v_y)f(v_z) = F(v_x^2 + v_y^2 + v_z^2)$$

where F is also an unknown function. It is possible to show that such equality is possible for an exponential function such that $f(v) = A\exp(\pm Bv^2)$. Maxwell chose the sign — since when v increases, the number of molecules must decrease. The final result is that the number of molecules with velocity between v and $v + dv$ is

$$NA^3 \exp(-Bv^2)4\pi v^2 dv$$

where $4\pi v^2 dv$ is element of volume in the velocity space between the spheres with radius v and $v+dv$. It remains to find the two constants A and B. It is done using the following data

$$N \int_0^\infty A^3 \exp(-Bv^2)4\pi v^2 dv = N$$

since the total sum is equal to the number of molecules. And the mean value of v^2 which is known to be equal to $(3RT/Nm)$ is given by

$$\int_0^\infty v^2 A^3 \exp(-Bv^2) 4\pi v^2 dv = (3RT/Nm)$$

The final result[2] is

$$f(v) = 4\pi(m/2\pi k_B T)^{3/2} v^2 \exp(-E/k_B T)$$

The importance of this work stands from the picture of a gas which is proposed: each molecules has only a certain probability to have a particular velocity since the expression $A^3 \exp(-Bv^2)4\pi v^2 dv$ is seen as a probability. It is the first introduction of probability in thermal physics. One cannot know the velocity of an individual molecules but only the probability that it has a given velocity. This step is conceptually very important since it opens the way to the development of statistical mechanics.

10.3 Boltzmann and Irreversibility

In the second half of the ninetieth century, explaining the existence of irreversibility by means of the laws of mechanics seemed to be impossible. The reason for that is that the equations of the mechanics are reversible with a change of the time from t to $-t$. We can take a very simple example of throwing a stone in the air with a velocity making some angle with the vertical direction. After a parabolic trajectory, the stone reaches the ground. But imagine that if it were possible to inverse the time after the throwing, we shall get the same trajectory than the first case but only with a change in the direction of the positions and of the velocities. The process of throwing is thus reversible. Applying this argument to the motions of molecules in a gas shows that the gas will never show irreversibility.

The proposal of Boltzmann to understand irreversibility can be formulated as follows. The state of a system is given by the positions and the velocities of all the components, say N molecules, i.e. by the

[2] We write $f(v)$ in the modern way. The Boltzmann constant was not known at the time of Maxwell.

knowledge of $6N$ quantities (3 for the position of one molecule and 3 for its linear moment if one sees these molecules as points). One can define a space phase with $6N$ dimensions and a "point" in this space represents a microscopic state of the system. This point will move constantly in this space. For the sake of the simplicity, this phase space is divided in cells in which the representative point is located.[3]

The basic idea of Boltzmann is that the different states of the systems are different probabilities to take place. The equilibrium macroscopic state has the largest probability. More precisely, the equilibrium macroscopic state corresponds to the largest number of cell points representing the system.

Since a macroscopic state has a certain probability, this does not exclude the possibility for other states different from the equilibrium state to exist. But their probability is very low such that it is not possible to observe them. A simple example of irreversibility is a drop of black ink falling in a water container. Just after the fall, one can distinguish colored and transparent regions of water. But after a while, the water takes a uniform grey color and one does not see states with separation of ink and water. Such states are not forbidden by the laws of mechanics but they have a very low probability to appear and the state of complete mixture has the largest probability. It is the equilibrium state. One sees the qualitative analogy with entropy. In the example of the ink and water, the entropy of the system (ink + water) increases until its largest value is reached at the equilibrium state. A direct relation must exist between the entropy and the probability of the equilibrium state.

It is the fundamental proposal of Boltzmann (in 1877) to relate first the probability of occurrence of a state to the number of cells in the phase space and secondly this probability to the entropy. The probability is merely proportional to the number W of points corresponding to a macroscopic state or the number of "complexions" realizing this state. The entropy is proportional to the logarithm of W or $S = k_B \,\mathrm{Ln}\, W$. The constant $\mathrm{k_B}$ is the famous Boltzmann constant. In equilibrium W and consequently S must be maximum.

[3]I shall not discuss the problem of the size of a cell and if it contains one or more representative points.

This proposal was not accepted immediately by numerous scientists and the debate about the Boltzmann formula was intense. Echoes of the discussions appeared long time after publication of the paper of Boltzmann.

The importance of this proposal must be emphasized since it is the first time that an interpretation of this mysterious quantity called entropy is given, which can only increase in an isolated system. This gives the possibility to relate microscopic and macroscopic quantities. Statistical mechanics was born.

Very often (in particular in the popular literature) the entropy is related to disorder. Microscopic systems are seen as disordered since the molecules have various and varying states (position and velocity). Frequently one says that W and S are measures of disorder. But doing that, one forgets the link with irreversibility and the macroscopic aspect of entropy.

10.4 Gibbs, the Father of Statistical Mechanics

The contribution of Gibbs to statistical mechanics can be found in a short treatise published in 1902, one year before his death: *Elementary Principles in Statistical Mechanics.* This *book* had a decisive influence of the development of statistical mechanics, term that Gibbs himself coined. The subtitle indicates the program of the book: *the rational foundation of thermodynamics.*

The goal of Gibbs is to develop rigorously what he calls a "broader view" of mechanics. He adopts the point of view of Maxwell in leaving out the possibility to follow in time the evolution of a mechanic system. He imagines an ensemble of identical systems varying by the configurations[4] and the velocities and he looks for the law giving the number of systems "which fall within any infinitesimal limits of configuration and velocity". It is clear that Gibbs knew the work of Boltzmann since he mentioned him on his preface. He used also largely the concept of space phase to describe states of a system.

Gibbs introduces the three ensembles: the canonical, the microcanonical and also the grand ensemble for system compound of

[4]By configuration Gibbs means the positions of the particles of the system.

several kinds of molecules. But contrarily to the method adopted in this book, he defines the canonical ensemble by a linear relation between the logarithm of the probability to find a given state and the energy: $\log P = (\psi - \varepsilon)/\theta$($\varepsilon$ is the energy, ψ and θ being constant) or $P = \exp[(\psi - \varepsilon)/\theta]$. This choice is justified only by its importance. Gibbs deduces all the properties of the canonical ensemble from his basic definition. He concludes that θ is related to the temperature and P to the entropy.

This small book gives us all the methods which are now standard in the study of thermal physics.

10.5 Planck and Einstein: Quantum Theory and Statistics

We put in parallel the work of Planck concerning the black body radiation and that of Einstein on the specific heat of solids. Both gave the correct interpretation of phenomena that the classical statistical mechanics was not able to explain. Both introduced the basic principle of quantum physics: the energy quanta.

At the end of the ninetieth century the radiation spectrum of a gas of photon was measured in large range of wave lengths and it was shown that this spectrum as a function of the frequency (or the wave length) exhibits a maximum. The classical interpretation begins with the equipartition theorem for one oscillator with energy $k_B T$. To get the radiation spectrum one multiples the energy of one oscillator by the frequency density of states of standing waves $(V/\pi^2 c^3)\omega^2 d\omega$ (ω is 2π the frequency v multiplied by 2π) and one has for the radiation spectrum the expression $[(\omega^2 k_B T)/(\pi^2 c^3)]d\omega$ which is correct only at low frequency. Thus the classical theory cannot explain the maximum in the spectrum.

The work of Planck was twofold. Examining the experimental results he found empirically the correct formula:

$$K(v, T) = [(8\pi h v^3)/c^3][\exp(hv/k_B T) - 1]^{-1}$$

The second step was to find the theoretical derivation of this formula that he did in 1900. He took the method of Boltzmann to calculate the entropy and from that the relation between energy and entropy he determined the temperature. The innovation in calculating W (the number of complexions) was to suppose that the energy of one oscillator was discontinuous by step of hv (h was a constant which was named the Planck constant and v is the oscillator frequency). This gives the possibility to get the formula already found only by inspection of the experiments. It is not clear if Planck was really conscious that his assumption will open the way toward a new of physics. It is likely that he was happy because "it works". One must also stress that from his work Planck was able to give the first determinations of his constant h and of the Boltzmann constant k_B.

I mentioned above that at the beginning of the ninetieth century Dulong and Petit made numerous measurements of specific heats solids until they formulated their law: the specific heat of solids is independent of T and equal to 6 cal/mole (approximately equal to $3R$). This law was thought so general that it was used to determine the atomic weight of some solids. However, there were exceptions. In particular measurements of carbon made in 1833 by Avogadro and in 1840 by de la Rive and Maret showed that, at ambient temperature, the specific heat of carbon is much lower than 6 cal/mole. In 1872 Weber succeeded in measuring the specific heat of diamond at low temperature (down to 20°K) and he observed a regular decrease of the specific heat when the temperature is lowered.

The theoretical interpretation of the Dulong and Petit law was given by Boltzmann in 1876 using the equipartition theorem for N three dimensional harmonic oscillators and he got the value of $3R$/mole.

In 1906 the puzzle of the specific heat at low temperature received its first solution by Einstein. He calculated the mean energy E_m of one harmonic oscillators by means of the formula of the canonical ensemble

$$E_m = \int_0^\infty E\exp(-E/k_BT)dE \Big/ \int_0^\infty \exp(-E/k_BT)dE$$

and this gives $E_m = k_BT$, the classical result. The idea to Einstein was to adopt the assumption of Planck concerning the energy of an oscillator. He supposed that the energy can take only discrete values by steps of hv. Putting $E = hv/k_BT$, he gets the mean energy

$$E_m = hv[\exp(hv/k_BT) - 1]^{-1}$$

For a solid Einstein supposes that all the atoms have the same frequency and this gives for the specific heat

$$C = 3Nk_B(hv/k_BT)^2 \exp(hv/k_BT)[\exp(hv/k_BT) - 1]^{-2}$$

Einstein compared his formula with data of diamond and found good agreement except at very low temperature. The assumption of independent motion of atoms is not correct and Debye gave later (in 1912) a method to calculate the specific heat taking into account the collective motions of the atoms. The works of Planck and Einstein were the first introduction of quantum ideas in statistical mechanics and their success gave strong evidence that a new physics is needed.

10.6 The Method of Bose and the Bose–Einstein Condensation

In 1924 Bose published a short paper in which he proposed a new derivation of the Planck formula. He wrote it first in English but the paper was refused by an English journal *The Philosophical Magazine.* He wrote to Einstein asking his opinion and Einstein translated him in German. Finally the paper appeared in *Zietschrift für Physik.* The novelty of the paper of Bose was to see the radiation gas as a gas of photons and to apply the relation $p = (hv)/c$. The density of states was the momentum density of states. He calculated the entropy by the number of complexions but supposing that the particles are undistinguishable.

The method of Bose was adopted by Einstein in considering molecules and not photons. He calculated W the number of complexions leaving out the possibility to distinguish between particles. He looked for a maximum of Ln W but with constraints, as Bose did: the number of particles is constant and the energy is also constant. The well known method of the multipliers of Lagrange permits to determine a maximum (or a minimum) of a function with constraints.

By this way, Einstein was able to get the well known formula giving the mean number of particles with energy E

$$N_E = [A^{-1} \exp(-E/k_BT) - 1]^{-1}$$

One recognizes in this formula the habitual expression if one take the multiplier A equal to $\exp(\mu/k_BT)(\mu$ is the chemical potential). In a paper in 1924 (shortly after the publication of the Bose paper) Einstein considered only the case $A < 1$ (or $\mu < 0$), i.e., a boson gas at not too high temperature. But later in 1925 he investigated the case $A = 1$ (or $\mu = 0$) and concluded that a growing number of molecules goes to the smallest energy level. Einstein established the condensation of a boson gas. The name of Bose is associated to that of Einstein since Einstein used the method first proposed by Bose. Today it is a standard method which is exposed in several textbooks but in this book I choose another more direct way.

10.7 The Principle of Pauli and the Statistics of Fermi and Dirac

The principle of Pauli is a strictly quantum effect. Pauli in 1925, looking for the interpretation of the electronic structure of atoms, proposed that two electrons cannot be found in the same quantum state. At the same time he saw the need for a new quantum number which is the spin.

The relation with statistics was seen independently by Fermi and by Dirac in 1926. In fact the first was Fermi who published a short paper in which he gave the expression of the Fermi–Dirac function. He followed a method parallel to that of Einstein for bosons maximizing the function Ln W with the two constraints of the fixed number of electrons and a fixed value of the energy. The determination of W was made in accordance with the Pauli principle.

Dirac found also the Fermi–Dirac function but he gave a more general view of the statistics in showing that there are two kinds of particles, bosons and fermions. Considering the wave function of an ensemble of particles, he showed that there are two possibilities when one makes permutation between the quantum states of two particles. Either there is no change of the wave function or there is a change in the sign. In the last case, it is possible to show that two particles cannot be in the same quantum state.

Interesting enough, this novelty in statistical mechanics was immediately appreciated and applied. For example at end of 1926, Ehrenfest and Uhlenbeck published an article on the two statistics giving their names of Bose–Einstein and Pauli–Fermi–Dirac. In 1927 Sommerfeld established the theory of electrons in metal using the Fermi–Dirac statistics.

Finally the relation with the spin (bosons with integer spin and fermions with half integer spin) was made independently by Fierz in 1939 and by Pauli in 1940.

Thus in 1940 all the tools to study statistically independent particles were present.

10.8 Modern Developments

To finish this brief history, I want to tell some words on recent advances. Thermodynamics is not a closed field and there are always new developments but at relatively slow rate. I want only to quote the names of Onsager and Prigogine who studied the thermodynamics of irreversible processes.

In statistical mechanics, emphasize was put on methods for studying interacting particles. The main field of application is condensed matter although there are also applications in other fields like astrophysics. The examples of applications are very numerous (semiconductors, superconductivity, magnetism, theory of liquids, transport phenomena, percolation etc). One very active research was phase transitions and critical points with the theory on renormalization (Wilson, 1972).

The final word will be for mentioning the application of statistical mechanics outside physics. There are many tentative to apply its methods to economy (with a new field, Econophysics), to geography, to social sciences (sociology, psychology) and probably others.

Chapter 11

Exercises

Chapter 1: The Microcanonical Ensemble

Exercise 1.1

A system is made of three identical harmonic oscillators in a closed box. The energy of one oscillator is given by:

$$E_i = \hbar\omega\left(n_i + \frac{1}{2}\right) \text{ when } i = 1, 2, 3$$

The n_i's are integers equal to $0, 1, 2, 3, \ldots$

The system is prepared such that the total energy is:

$$E = \sum E_i = \hbar\omega\left(n + \frac{3}{2}\right) \text{ when } n = n_1 + n_2 + n_3$$

(a) Suppose that $n = 3$, find the entropy of the system.
(b) Suppose that $n = 5$, find the entropy of the system.
(c) Now n is chosen to be a not specified integer, calculate the entropy of the system as a function of n, the temperature T and the energy $E(T)$ when $n \gg 1$.

Exercise 1.2

A closed system is made of two subsystems A and B in thermal contact one with another. In each subsystem there are N particles with different possible energies. The particles in the subsystem A can have two possible energies: 0 and ε and in the subsystem B they can have two possible energies: ε and 2ε. The total energy E which is the sum of the energies E_A and E_B is known. Use the microcanonical ensemble.

(a) Calculate the energies E_A and E_B of each subsystem. Express your answers with the help of E, N and ε.
(b) Calculate the temperature of the system.

Exercise 1.3

One considers N identical harmonic oscillator, and the total energy is given by

$$E = \sum E_i = \hbar\omega\left(n + \frac{3N}{2}\right)\ (i = 1, 2, 3, \ldots, N)$$

The numbers N and n are very large numbers such you can use the Stirling formula:

$$\mathrm{Ln}\, N! = N\,\mathrm{Ln}\, N - N$$

Calculate the entropy as function of n and N, the temperature T and the energy E as a function of T.

Hint: You have to calculate the number of possibilities to put n units, say "particles", in N "boxes" or oscillators. Suppose that we put the n particles on a line and also put $N - 1$ walls between the particles giving a picture in N boxes. You have to find now all the complexion of this $n + N - 1$ objects (n particles $+N - 1$ walls) taking into account that the particles are indistinguishable and the walls also.

Exercise 1.4

The entropy of magnetic material is given by

$$S = S_0 + \frac{1}{2}A\left(\frac{H}{T}\right)^2$$

where T is the temperature, H is the magnetic field and A is a constant.

(a) What is the energy as a function of H and T?
(b) What is the magnetization as a function of T and H?
(c) Show that the above results correspond to a material with one magnetic dipole b for each of the N atoms, at high temperatures. The magnetic dipoles can have two directions $+x$ and $-x$ and the field is directed along $+x$. Use the microcanonical ensemble and determine the constant A.

Hint: We suggest the following procedure. First, calculate the energy, and from the limit of the energy at high temperatures, determine the entropy.

Chapter 2: The Canonical and Grand Canonical Ensembles

Exercise 2.1

Solve the problem 1.2 by mean of the partition function.

Exercise 2.2

N particles are in thermal contact with a reservoir at temperature T. The possible energies of the particles are as follows:

Energy $E_1 = 0$ and degeneracy 1 (only one state with energy E_1)
Energy $E_2 = \varepsilon$ and degeneracy 2 (two states with energy E_2)
Energy $E_3 = 2\varepsilon$ and degeneracy 1 (one state with energy E_3)

(a) Calculate the one particle partition function and the total partition function.
(b) Calculate the energy of these N particles and find the limits for $T \to 0$ and for $T \to \infty$.
(c) Calculate the entropy and the limits for $T \to 0$ and for $T \to \infty$.

Exercise 2.3

N particles have p possible states with energies E_i (i from 1 to p). Show that in the limit of high temperatures, the entropy tends toward:

$$S(T \to \infty) = N k_B \operatorname{Ln}(p)$$

Exercise 2.4

N particles have two possible energies $E_1 = 0$ and $E_2 = e$ and they are in contact with a reservoir at temperature T. The walls of the reservoir are permeable to particles.

(a) Give an expression of the grand partition function as an infinite series.
(b) At what condition the series will converge? Express the condition with the help of the chemical potential, T and e.
(c) Suppose that the series of the grand partition function converges. Calculate N, the mean number of particles with the help of the chemical potential, T and e.
(d) Find the chemical potential as a function of N, T and e. Calculate the energy as function of N, T and e.
(e) Check if the condition of convergence found in (b) is verified.

Exercise 2.5

The entropy of magnetic material is given by

$$S = S_0 + \frac{1}{2} A \left(\frac{H}{T}\right)^2$$

where T is the temperature, H is the magnetic field and A is a constant.

(a) What is the energy as a function of H and T?
(b) What is the magnetization as a function of T and H?
(c) Show that the above results correspond to a material with one magnetic dipole for each of the N atoms, at low temperatures. The magnetic dipoles can have two directions $+x$ and $-x$ and the field is directed along $+x$. Use the canonical ensemble.

Exercise 2.6

One considers N particles with two energy levels 0 and e. They are in contact with a reservoir at temperature T.

(a) Show that the specific heat C goes to zero for high temperatures.
(b) Propose an explanation for this behavior of C.

Exercise 2.7

N particles are in contact with a reservoir at temperature T. Their energy is given by $E = n_i e_0$ and $e_0 = A(V/V_0)^\alpha$ where V is the volume of the particles and V_0, A and α are constants.

(a) Calculate the partition function.
(b) Calculate the equation of state $P(V, T)$. P is the pressure.
(c) Calculate the entropy $S(V, T)$ and the limits at low and high temperatures.
(d) What are the limits of the constant α?

Exercise 2.8

One considers an ideal gas with N particles. Its volume is V and its temperature T. One wants to know what the differences are if the particles are seen as distinguishable or not. For that one calculates the properties of the gas in these two cases:

Case 1: Distinguishable particles
Case 2: Nondistinguishable particles

(a) Write the partition function in the two cases. You are not asked to calculate them but only to copy from the textbook.
(b) Calculate in the two cases the following quantities: the energy E, the state function $P(V, T)$, the free energy F and the entropy S. What are the functions which are identical and those which are different?
(c) Verify if there is a temperature interval for which there are no differences between the two cases.

Exercise 2.9

N particles at temperature T have energy levels given by $E = n_1 e_i$ when the numbers n_i goes from 1 to Q.

(a) Calculate the energy E and the entropy S.
(b) What are the limits of E and S when T goes to large values?

Hint: Use the following approximation for $\exp(x)$ when $x \to 0$:

$$\exp(x) = 1 + x + \frac{x^2}{2}$$

Exercise 2.10

This problem investigates the changes in the properties of an ideal gas when the number of particles varies. For that, one considers a quasi-static process at constant entropy and volume and increasing the number of particles which is N_1 at the beginning of the process. The initial temperature is T_1

(a) Show that in such a process, one has $\left(\frac{dN}{dT}\right)_{S,V} < 0$.
(b) Show that in this process the energy is decreased if the number of particles of an ideal gas is increased.

Chapter 3: Quantum Statistics

Exercise 3.1

Show that in a system of quantum particles with energies e_i the total energy is

$$E = \sum e_i \left(\frac{\partial F}{\partial e_i}\right)$$

where F is the Helmotz free energy.

Exercise 3.2

The possible energies of N quantum particles are $E_1 < E_2 < E_3 < E_4$ with degeneracy $G_i (i = 1, 2, 3, 4)$.

(a) Sketch a qualitative graph of the chemical potential as a function of the temperature in the following cases:

1. The particles are bosons
2. The particles are fermions with $N = G_1 + G_2$

Plot the two curves on the same graph.

(b) Give an expression of the energy in the classical limit (Maxwell–Boltzmann statistics) and find the energy at very high temperatures.

Exercise 3.3

(a) Find the relationship between n_1 the mean number of quantum particles in the lowest level (energy E_1) and the mean number n_i in one higher energy level E_i.

1. In the case of bosons
2. In the case of fermions

(b) Calculate n_i as a function of n_1, the temperature T and the difference $E_i - E_1$.

(c) What are the limits of n_i at low temperatures and at high temperatures, for bosons and for fermions?

Exercise 3.4

Particles without fixed number can have only two possible energies: ε_1 and $\varepsilon_2 > \varepsilon_1$. The two levels have the same degeneracy G(G states for each energy)

(a) The particles are bosons. Calculate the energy $E(T)$, the specific heat $C(T)$ and the mean number of particles $N_1(T)$ and $N_2(T)$ in the two levels. Give for these three quantities the limits for $T \to 0$ and $T \to \infty$.

(b) Same questions if the particles are fermions.

(c) Plot the specific heats of bosons and fermions on the same graph.

(d) Plot the ratios N_2/N_1 as a function of T for bosons and fermions on the same graph.

Hint: The graphs are only qualitative.

Exercise 3.5

Two particles have three possible energies $0, e, 2e$. They are in contact with a reservoir at temperature T.

(a) What are the different configurations if the two particles are two fermions with spin 1/2.
(b) Give an expression of the energies in each of the configurations and calculate them in the limit $T \to \infty$.
(c) Same questions if the two particles are bosons with spin 0.

Exercise 3.6

N fermions are at temperature T. For each particle there are two possible energies 0 and e. The degeneracy of each energy level is equal to $M \geq N$.

(a) Find the equation from which it is possible to calculate the quantity $\exp(-\beta\mu)$ where $\beta = kT$ and μ is the chemical potential.
(b) Solve this equation for $N = M$ and calculate the energy of each level.
(c) Calculate the number of particles in each energy level.

Exercise 3.7

Same problem than the precedent but for bosons.

Chapters 4 and 5: Miscellaneous

Exercise 4.1

(a) Show that the density of states of particles with linear moment p located on a surface of area A (particles in two dimensions, 2D) is $g_2(p)dp = (2\pi A)h^{-2}pdp$.
(b) Show that the density of states of particles with linear momentum p located on a line of length L (particles in one dimension, 1D) is $g_1(p)dp = \left(\frac{2L}{h}\right) dp$.

Exercise 4.2

Calculate the energy density of states $g(E)dE$ of particles for which the energy and the momentum are related as $E = Kp^s (s > 0)$ in 3D, 2D and 1D.

Exercise 4.3

N atoms of an ideal gas with mass m and linear momentum p are located in a cubic box with edge L. One side of the box can absorb atoms and these atoms behave as an ideal gas in two dimensions.

(a) Calculate the partition function of the ideal gas on the absorbing side if one supposes that there are N_S absorbed atoms on this side.
(b) Calculate the grand partition function of the atoms inside the box and the grand partition function of the absorbed atoms.
(c) Calculate the number of atoms N_V in the box and the number N_S of atoms onto the absorbing side. Give the ratio (N_S/N_V).

   1. By means of the partition functions
   2. By means of the grand partition functions

Exercise 4.4

A gas of quantum particles (fermions or bosons) has the following properties:

1. The possible energies form a continuum from 0 to infinity $(E \geq 0)$
2. The energy density of states is $g(E)dE = AE^{\alpha}dE$

(a) Demonstrate the following relationship: $PV = KE$ where K is a function of α (P is the pressure and V the volume).

   Hint: Transform the integral giving $\ln Z_G$ by integration by parts.
(b) Apply the results to the cases of relativistic particles $(E = pc)$ and nonrelativistic massive particles $(E = p^2/2m)$.
(c) Give the equation of state in the two preceding cases.

Exercise 4.5

An ideal gas of N atoms have their energy related to the momentum as $E = pc$ (c light velocity), The volume is V and the temperature T.

(a) Calculate the one atom partition function.
(b) Calculate the energy and the specific heat at constant volume.
(c) Calculate the equation of state.

Exercise 4.6

Find the energy and the entropy as functions of the temperature T of N classical harmonic oscillators for which the energy of one oscillator is $E = p^2/2m + Kx^2/2$ (p is the momentum, m the mass and K the spring constant).

Calculate the chemical potential of these oscillators.

Exercise 4.7

A linear harmonic oscillator has a magnetic moment dependent on its position. The energy of one oscillator is $E = p^2/2m + Kx^2/2 - \gamma Hx$ (p is the momentum, m its mass, K the spring constant, H the magnetic field and γ is a constant).

(a) Calculate the one oscillator partition function in classical mechanics.

    Hint: Use the identity $Ax^2 - Bx = A(x - B/2A)^2 - B^2/4A$
(b) Calculate the energy of N identical oscillators as a function of T and H.
(c) Calculate the magnetization of N oscillators.

Exercise 4.8

N particles have energy of infinite levels given by $E_i = Pn_i$ where P is a const and $n_i = 1, 2, 3, \ldots$ The constant P is small such that one can consider the energies as a continuum, $E = Ke$. Calculate the chemical potential μ.

(a) In the case of bosons.
(b) In the case of fermions. For what temperature T_0 μ is null?

Give also the expressions of μ for $T \to 0$ and for $T \to \infty$ for the bosons and the fermions

Hint: $\int_0^\infty \frac{1}{a[\exp(x)+1]} = \text{Ln}\left(\frac{1}{a}+1\right) \int_0^\infty \frac{1}{a[\exp(x)-1]} = \text{Ln}\left(1-\frac{1}{a}\right)$

Exercise 4.9

One asks to find the energy at low temperature of the N particles of the preceding problem. You are asked to find the energies in the following cases:

(a) Bosons. Take as an approximation $\mu = -k_B T$.
(b) Fermions. Take as an approximation $\mu = M - k_B T$.

Hint: You do not need to calculate the integrals but transform them in definite integrals which are constant. In this problem, one looks only for the temperature dependence.

Chapter 6: Gas of Photons

Exercise 6.1

(a) Calculate the thermal properties $E(T), S(T), C(T)(C(T)$ is the specific heat at constant area) of a gas of photons located onto a surface of area A. Give the emission spectrum $K(\lambda)$.
(b) Same problem for a gas of photons in 1D, on a line of length L.

Exercise 6.2

Give the expression of the emission spectrum of a photon gas in a volume V as a function of the frequency ω. Give the limits of the spectrum for $(\hbar\omega/k_B T) \ll 1$ and $(\hbar\omega/k_B T) \gg 1$.

Exercise 6.3

Calculate the emission spectrum of a gas of fermions in not fixed number (volume V, temperature T). Express the spectrum as a function of the wavelength and as a function of the frequency. Give the limits of the spectrum for $(\hbar\omega/k_B T) \ll 1$ and for $(\hbar\omega/k_B T) \gg 1$. Compare with the case of photons.

Exercise 6.4

Find the equation giving the relationship between the frequency and the maximum of the spectrum.

Try to find an approximated solution.

Exercise 6.5

A gas of photons is trapped in a cavity of volume V_1 and temperature T_1. By an adiabatic process one reduces the volume of the cavity to $V_2(V_2 < V_1)$.

(a) What is the relationship between the pressure and the volume?
(b) What is the new energy?
(c) By an isothermal process, one reduces the volume of the cavity to $V_2(V_2 < V_1)$. What is the new energy of the gas?

It exchanges heat with the exterior world. Does the gas receive heat or send heat outside?

Exercise 6.6

A Carnot cycle is made with a gas of photons. One begins by a state with a volume V_1 and pressure P_1. The first step is to make an isothermal expansion in contact with a hot reservoir at temperature T_{in} until volume $V_2 > V_1$ and $P_2 < P_1$. The second step is an adiabatic expansion until $V_3 > V_2$ and $P_3 < P_2$. The third is an isothermal compression in contact with a cold reservoir at temperature T_{out} until $V_4 < V_3$ and $P_4 > P_3$. The final step is compression until the starting point, what is the efficiency of such a cycle?

Exercise 6.7

What is the ratio between the specific heat and the mean numbers of photons in a photon gas at temperature T and volume V. Give your results in 1D, 2D and 3D.

Chapter 7: Phonons

Exercise 7.1

N atoms are located onto a surface forming a 2D solid of area A. Calculate the thermal properties: the energy $E(T)$ and the specific heat $C_V(T)$ following the Debye method. Give the limits for the low temperature and for the high temperature.

Exercise 7.2

Same question for N atoms along a line of length L.

Exercise 7.3

Calculate the mean numbers of phonons in media at 1D, 2D and 3D. Use model of Debye and give explicit expressions at low and high temperatures. Compare with photons.

Exercise 7.4

N atoms in a solid medium are in a volume V and in contact with reservoir at temperature T. Calculate the pressure of this system in the following cases at low temperature:

(a) Atoms connected by springs.
(b) The model of Debye.

Hint: The pressure $P = -\frac{\partial F}{\partial V}$ (F is the free energy). But at low temperature it is possible to use the approximate $P = -\frac{\partial E}{\partial V}$.

We propose not to use the Debye temperature in the case of the Debye model but only the maximum energy which depends on V.

Exercise 7.5

The models of the linear chain adapted at 3D and the Debye model contain very different elements. Nevertheless, at low and high temperatures they give very similar results. Can you explain why? Do not give answer with the help of formula but only physical arguments.

Exercise 7.6

Do you think that it is possible to define a Debye temperature for liquids? If yes, what does happen at melting from the solid state. The Debye temperature will increase or decrease?

Hint: The sound velocity of liquids is typically equal to 1500 m/s and to 4000 m/s for metals.

Exercise 7.7

A linear atomic chain is compound of N atoms with mass m and connected by springs of constant B. The linear motions of the atoms are longitudinal waves and one recalls the dispersion relation between the frequency $\omega\omega$ and the wave vector k:

$$\omega = 2(B/m)^{1/2}\sin(ka/2)$$

(a) Calculate the sound velocity as a function of B, m and the inter-atomic distance a.
(b) Give another expression of the sound velocity as a function of the largest possible frequency.

Exercise 7.8

One considers a linear chain made of atoms with mass m. They are connected by springs with constant B. One wants to determine the transversal motion of the atoms. The displacement of the atom i is v_i perpendicular to the wave vector (propagation in the direction x perpendicular to y). Give the dynamic equation for the displacements and the relation $\omega(k)$.

Exercise 7.9

In a linear chain of N identical atoms of mass m the atoms are connected by springs with constant B. One wants to know the thermodynamic of this chain taking into account longitudinal phonons and transversal phonons.

(a) Give an expression for the energy $E(T)$ and the limits for low and high temperature.
(b) Give also an expression for the specific heat and the limits at low and high temperature.

Chapter 8: Gas of Bosons

Exercise 8.1

A gas of N bosons with mass m is located onto a surface with area A.

(a) Write the equation from which it is possible to calculate the Bose–Einstein condensation temperature if it takes place. Show that the Bose–Einstein condensation does not take place.

Hint: You have two possibilities to answer. One is to calculate the integral knowing that $\int dx[\exp(x) - 1]^{-1} = \text{Ln}[1 - \exp(-x)]$. In the second, you do not need to calculate the integral. Try the two methods.
(b) Calculate the chemical potential μ as a function of T and find the limits for the low temperatures and the high temperatures.

Hint: $\int dx[C\exp(x) - 1]^{-1} = -\text{Ln}[1 - C^{-1}\exp(-x)](C > 1)$
(c) Show that $(d\mu/dT)$ goes to zero with T. Sketch a qualitative graph of $\mu(T)$.
(d) Show that the ratio (μ/k_BT) goes to zero with T.
(e) Calculate the energy at low temperatures.

Exercise 8.2

The dimension of the space can be only 3, 2 or 1. However, suppose that there are spaces with a non-integer dimension d. From the results of the exercise 4.2 one can write the momentum density of state as

$$g(p)dp = Ap^{d-1}dp$$

where d is the dimension of the space. Find for what dimensions between 3 and 2 one can observe a Bose–Einstein condensation.

Hint: You do not need to calculate any integral to answer this question.

Exercise 8.3

N bosons have the following properties:

1. They are located in a volume V
2. The energy E of one boson is equal to pc (p is the momentum, c light velocity)

(a) Calculate the Bose–Einstein condensation temperature.
(b) Calculate the ratio N_1/N (N_1 is the number of bosons in the lowest level) as a function of the temperature in the condensed regime.
(c) Calculate the energy $E(T)$ in the condensed regime.

Hint: $\int_0^\infty x^2[\exp(x) - 1]^{-1} \approx 2.404 \int_0^\infty x^3[\exp(x) - 1]^{-1} \approx 6.5$.

Exercise 8.4

The compressibility of a fluid is defined as $\chi = \frac{1}{V}\frac{\partial V}{\partial P}$ at entropy constant or at temperature constant, for a gas of N bosons (mass m) in a volume V and in contact with a thermal reservoir. The gas is in a state of partial condensation.

(a) Calculate the compressibility at constant entropy, supposing the gas remains partially condensed.
(b) Try to calculate the compressibility at constant temperature.

Exercise 8.5

(a) What is the relation between V and T during an isentropic (constant entropy) process between two temperatures smaller than the condensation temperature for a boson gas of N atoms?
(b) What is the relation between the pressure and the volume for the same process than in (a)?

Exercise 8.6

The velocity of the sound in a fluid is given by

$$V = \sqrt{\frac{1}{\chi\rho}}$$

where χ is isentropic compressibility and ρ is the density.

(a) Calculate the velocity of sound of a gas of N bosons at temperature T smaller than the condensation temperature (mass m) and volume V.
(b) Make a numerical evaluation of the velocity for a helium atom made of two protons and two neutrons.

Exercise 8.7

A gas of N bosons is in a volume V and in contact with a reservoir at temperature T. During a process at constant temperature the volume is decreased from V_1 to V_2. The gas remains in a state of partially condensed.

(a) If the number of bosons is N_1 at the beginning of the process what is the number of bosons in the condensed state at the end of the process?
(b) What is the change in its internal energy?
(c) What is the variation of calorific and mechanical energies during this isothermal process?

Exercise 8.8

(a) Verify the relation for a gas of bosons in partial state of condensation $PV = KE$ where E is the energy. Determine the value of the coefficient K.
(b) Adapt the formula for a gas at 2D and 1D.

Exercise 8.9

A system is made of two boxes and each of them contains a gas of bosons. Their temperature T is the same for the two systems and also their condensation temperature $T_0 > T$. The total energy E is known. The number N_1, the volume V_1 and the energy E_1 of the subsystem 1 are known.

(a) Calculate the condensation temperature.
(b) Calculate the energy E_2 of the subsystem 2.
(c) Calculate the number N_2 and the volume V_2 of the subsystem 2.

Chapter 9: Gas of Fermions

Exercise 9.1

A gas of N fermions is located in a volume V at temperature T.

(a) Calculate the pressure of this gas at $T = 0$ and $T \neq 0$ (but low).
(b) Give an expression of the energy libre a $T > 0$.

Hint: Take the expressions of energy $E(T)$ as given in Chapter 9, and that of the Fermi level at $T > 0$.

Exercise 9.2

(a) Calculate the entropy of the gas of the precedent exercise.
(b) What is the relation between the volume and the temperature during a process at constant entropy?

Exercise 9.3

For N electrons in a volume V at temperature T the energy E is related to the linear momentum p as $E = pc$ (c is the light velocity).

(a) Calculate the Fermi level and the energy for $T = 0$.
(b) Show that for $T \neq 0$ (but not too high) the specific heat at constant volume is linear with T.

Exercise 9.4

(a) Show that in a process at constant entropy and constant volume, the energy is increased if the number of particles of this gas is increased.
(b) Try to give a physical explication for this result, without calculation.

Exercise 9.5

The energy spectrum of N electrons in a volume V is as follows:

1. There is one level (energy $E = -\varepsilon$) with degeneracy N (one state for one electron)
2. For energy equal or larger than 0 there is a continuum of energy up to infinity

(a) Find the position of the Fermi level at $T = 0$ (without calculation).
(b) Give an expression for the number n_1 of the free electrons (as a function of T, N, ε and the Fermi level μ).
(c) Give an expression for the number of electrons n_2 on the level with energy $-\varepsilon$ (as a function of T, N, ε and the Fermi level μ).
(d) Give an expression of the missing places $N - n_2$ on the level with energy $-\varepsilon$.
(e) Find the Fermi level as a function of T (for low T) and calculate n_1 as a function of T.

Hint: Find the approximate expression for $N - n_2$ taking into account the position of the Fermi level.

Exercise 9.6

A gas of electrons is located onto a surface of area A (electron gas in two dimensions).

(a) Calculate the Fermi level μ as a function of the temperature T. For what temperature it becomes null? What is its value at $T = 0$?

Hint: $\int dx[C \exp(x) + 1]^{-1} = -\mathrm{Ln}[1 + C^{-1} \exp(-x)]$.

(b) Calculate the derivative $(d\mu/dT)$ as a function of T and give its value for $T = 0$. Sketch a qualitative graph of $\mu(T)$.
(c) Show that for $T \neq 0$ (but not too high) C_V the specific heat at constant volume is linear with T.
(d) For a 2D boson gas at low temperatures the constant volume specific heat is also linear with T. Can you find an explanation why C_V has the same temperature dependence for bosons and fermions?

Exercise 9.7

(a) Calculate the relation between the volume (or the radius) and the mass of a neutron star: Radius R, mass M, volume V.
(b) If you think that the result is a paradox, try to find an explanation.

Exercise 9.8

A white dwarf is a star made of some atomic nucleons which are ionized. It is compound of these nucleons and free electrons. It is generally admitted the properties of a white dwarf are, in first approximation due to the electrons.

(a) Calculate the electronic kinetic energy of such a star admitting that the temperature is zero and the electrons are relativistic: it is the energy of Fermi $E_F(E = pc)$.
(b) Calculate the typical radius of a white dwarf following the schema of Chapter 9. First consider the total energy: Energy of Fermic plus energy of gravitation $U = -\frac{3}{5}\frac{GM^2}{R}$ (M is the mass of the nucleons). Secondly determine the volume of the star by minimizing the total energy. $E_F + U$ relatively to the volume V.
(c) Make a numerical evaluation of the radius.

Data:

Mass of the star = mass of the sun, $M = 2 \cdot 10^{30}$ kg
Gravitation constant $G = 6.7 \cdot 10^{-11}\,MKS$
Electron mass $m = 10^{-30}$ kg
Planck constant $h = 6.6 \cdot 10^{-34}\,MKS$

Exercise 9.9

Show that at the equilibrium, one has the following relation for a neutron star

The total energy $E = -\mathrm{U}$ where U is the gravitational energy

Index

www.ingramcontent.com/pod-product-compliance
Ingram Content Group UK Ltd.
Pitfield, Milton Keynes, MK11 3LW, UK
UKHW021827091026
9241IPUK00004B/69

9 789819 814343